Contentious Histories of Law, Feminism, and Forensic Science

In 1984, the Sexual Assault Evidence Kit (SAEK) was dubbed "Ontario's most successful rapist trap." Since then, the kit has become the key source of evidence in the investigation and prosecution of sexual assault as well as a symbol of victims' improved access to care and justice. Unfortunately, the SAEK has failed to live up to these promises.

The *Technoscientific Witness of Rape* is the first book to chart the thirty-year history of the Ontario Sexual Assault Evidence Kit and its role in a criminal justice system that re-victimizes many assault victims in their quest for medical treatment and justice. Drawing on actor-network theory and feminist technology studies, Andrea Quinlan analyses sixty-two interviews with police, nurses, scientists, and lawyers, as well as archival records and legal cases, to trace changes in sexual assault forensics, law, advocacy, and anti-violence activism in Ontario. Through this history Quinlan bravely and provocatively argues that the SAEK reflects and reinforces the criminal justice system's distrust of sexual assault victims.

ANDREA QUINLAN is an assistant professor in the Department of Gender and Women's Studies at Trent University.

ANDREA QUINLAN

The Technoscientific Witness of Rape

Contentious Histories of Law, Feminism, and Forensic Science

UNIVERSITY OF TORONTO PRESS
Toronto Buffalo London

© University of Toronto Press 2017
Toronto Buffalo London
www.utppublishing.com
Printed in Canada

ISBN 978-1-4875-0081-8 (cloth)   ISBN 978-1-4875-2060-1 (paper)

♾ Printed on acid-free, 100% post-consumer recycled paper with vegetable-based inks.

Library and Archives Canada Cataloguing in Publication

Quinlan, Andrea, 1984–, author
The technoscientific witness of rape : contentious histories
of law, feminism, and forensic science / Andrea Quinlan.

Includes bibliographical references and index.
ISBN 978-1-4875-0081-8 (cloth). – ISBN 978-1-4875-2060-1 (paper)

1. Rape – Investigation – Ontario – History – 20th century.
2. Rape – Law and legislation – Ontario – History – 20th century.
3. Rape victims – Legal status, laws, etc. – Ontario – History – 20th
century.   4. Evidence preservation – Ontario – History – 20th century.
5. Evidence, Criminal – Ontario – History – 20th century.   6. Forensic
sciences – Ontario – History – 20th century.   7. Criminal justice,
Administration of – Ontario – History – 20th century.   I. Title.

HV8079.R35Q56 2017   363.25'953209713   C2016-908257-1

This book has been published with the help of a grant from the Federation
for the Humanities and Social Sciences, through the Awards to Scholarly
Publications Program, using funds provided by the Social Sciences and
Humanities Research Council of Canada.

University of Toronto Press acknowledges the financial assistance to its
publishing program of the Canada Council for the Arts and the Ontario
Arts Council, an agency of the Government of Ontario.

Canada Council   Conseil des Arts
for the Arts   du Canada

ONTARIO ARTS COUNCIL
CONSEIL DES ARTS DE L'ONTARIO
an Ontario government agency
un organisme du gouvernement de l'Ontario

Funded by the   Financé par le
Government   gouvernement
of Canada   du Canada

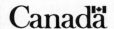

Contents

Figures

Acknowledgments

There are many people to whom thanks are due. To the advocates, activists, and medical and legal professionals who participated in this study and generously shared their memories and reflections on the SAEK's past, I extend sincere thanks. Your contributions brought this book to life. I am also greatly indebted to the many archivists at Library and Archives Canada, Archives of Ontario, City of Toronto Archives, University of Ottawa Archives and Special Collections, Clara Thomas Archives and Special Collections, and Miss Margaret Robins Archives of Women's College Hospital, who dedicated their time and wisdom to this project.

The research for *The Technoscientific Witness of Rape* was supported by the Social Sciences and Humanities Research Council of Canada and York University. I am grateful to many people who provided feedback on earlier versions of the manuscript including Patricia McDermott, Soren Frederiksen, Meg Luxton, Annette Burfoot, Amanda Glasbeek, and Michael Lynch. I am also thankful to the Department of Science and Technology Studies at Cornell University and the Center for Science, Technology, Medicine, and Society at the University of California, Berkeley for providing me with inspiring institutional homes while I wrote and revised this manuscript. Thank you to the staff at the University of Toronto Press, in particular Douglas Hildebrand, who expertly guided this book through to publication and to the anonymous reviewers who generously provided detailed comments and valuable suggestions on the manuscript. I am especially grateful to my teachers: Janice Newson, for her insight, steady guidance, and friendship, and Arthur

Frank, for his inspiration, mentorship, and encouragement to always take non-human actors seriously.

And finally, for their love, unwavering support, and patience, a very special thank you goes to Elizabeth Quinlan, Michael Rohatynsky, Kevin Quinlan, Lukin Robinson, and Curtis Fogel.

My number was called and I'm wheeled into a medium-sized room where three women in white lab coats are waiting. They speak softly, tell me their nurse and doctor names, [and] what they will do ... Like an annual physical, they light my eyes, touch me, bend, tap, prod and weigh me. They ask if I take medication, have attempted suicide, been admitted to a psychiatric hospital. If I have had children, abortions, recent consensual sex ... then they ask me to tell my rape. They write everything down, record the data on forms with numbers and codes that have been waiting for me to be raped. All of this is standard procedure. The women who treat me explain everything in apologetic tones, whisper commands in powdered voices – things better not spoken too loud. They are efficient and distant as they spread a circular plastic sheet on the floor and ask me to stand on it and remove my robe. They brush the hair on my head and between my legs pluck[ing] fifteen pubic hairs by the root ... It should hurt but I don't feel it. I don't feel anything ... stirrups, gloves, stainless steel inside me, entering, expanding. From a nearby microscope a woman's voice says, "We've got sperm here, one's still alive." I tell her to kill it and she looks at me and smiles. Everything is collected in vials and plastic or under glass, labelled with my name. All these pieces of me are placed in a kit to be touched and examined, probed and considered some more, somewhere, by someone, for something.

Jane Doe, 2003, 13

THE TECHNOSCIENTIFIC WITNESS OF RAPE

Contentious Histories of Law, Feminism, and Forensic Science

1 Introduction: Diffracting the Technoscientific Witness

In 1977, the Women's Press published the first Canadian book on rape. Frustrated with the lack of Canadian data on rape, the authors, Lorenne Clark and Debra Lewis, had set out to break new empirical ground. And indeed they did. Through a detailed analysis of rape incident reports at the Metropolitan Toronto Police Department, Clark and Lewis revealed how the criminal justice system was systematically failing to live up to its promise of protecting women from rape. They charted the alarming rates at which police were dismissing women's rape reports as untruthful and unfounded, as well as the small fraction of rape cases that resulted in conviction. Their analysis revealed a paradox of law:

> At the level of codified law and public pronouncements, we repudiate rape as a serious offence ... But at the level of actual practice, women have found little real protection in the judicial system. Few cases come to court, fewer rapists are convicted, and the victim, rather than the rapist, is put on trial. Our courts do not fulfill the promise of the law, and rape remains a serious threat to women. (Clark & Lewis, 1977, 24)

In the years after its publication, Clark and Lewis's book, *Rape: The Price of Coercive Sexuality*, took on a life of its own as activists and advocates used it to draw attention to women's ongoing plight in the legal system and the failures of police and the courts to take rape seriously.

The book was written amidst the feminist anti-rape movement of the 1970s. Two years before Clark and Lewis's book was released, American radical feminist Susan Brownmiller had written *Against Our Will: Men, Women, and Rape*, which provided one of the first comprehensive analyses of male rape against women. Rape is a weapon of power that men employ to keep all women in fear, Brownmiller had provocatively argued. Her words became a well-cited axiom in radical feminist circles. Although on different sides of the Canadian-US border, Brownmiller, Clark, and Lewis were part of a growing movement that was bringing rape into public consciousness.

Riding the momentum of second wave feminist activism on gender inequality and sexism in workplaces, labour unions, and families, feminist anti-rape activists sought to expose the commonality of male violence against women and the systemic failures of medical and legal institutions to respond. Activists in Canada and the United States worked to fill the void of medical services and legal protections for rape victims. They organized community-based rape crisis centres to provide advocacy, counselling, and support for victims and their allies. They built feminist expertise in victim advocacy and trained volunteer advocates to support and accompany victims to hospitals, police stations, and courtrooms. Anti-rape activists pushed medical and legal institutions to reform their laws and practices that activists saw as patriarchal, oppressive, and rarely serving the interests of women who were raped. They lobbied government agencies, rallied in the streets, and pressed police and hospitals to hear their calls for improved institutional practice. In Canada, these pressures inspired rape law reforms, specialized sexual assault training for nurses and police, and standardized protocols for forensic evidence collection.

Forty years later, there is now an elaborate network of professionals, expert practices, and forensic technologies dedicated to treating, investigating, and prosecuting sexual assault. In this network, standardized sexual assault evidence kits are now used to collect and store forensic evidence from victims' bodies. These kits, and the socio-technical network they operate within, are heralded by some as significantly improving victims' access to care and justice. Standardized protocols for forensic evidence collection, according to some medical practitioners, have vastly improved victims' treatment in hospitals and have increased the likelihood that their case will end in conviction (Aiken & Speck, 1995; Campbell, Patterson, & Fehler-Cabral, 2010; Campbell, Patterson, & Bybee, 2012; Campbell, Patterson, Bybee, & Dworkin, 2009; Cornell, 1998; Little, 2001). Specialized training for nurses and police, according to others, has considerably improved victims' treatment and care (Campbell, Patterson, & Lichty, 2005; Dandino-Abbott, 1999; Stermac & Stirpe, 2002; Plichta, Clements, & Houseman, 2007). Although these praises of contemporary practice might suggest that the picture Clark and Lewis painted in 1977 is no longer accurate, there is much evidence to the contrary.

The legal and institutional reforms around sexual assault that anti-rape activists began fighting for in the 1970s have not lived up to their promises. In fact, despite decades of activism and resulting institutional reforms, there is mounting evidence suggesting that little has changed for victims in the criminal justice system (e.g., Bumiller, 2008; Corrigan, 2013a; Doe, 2012; Martin, 2005; Mulla, 2014). Victims continue to report feeling dismissed, doubted, and re-traumatized in the hospitals, police stations, and courtrooms to which they turn for help (Mulla, 2014; Ullman, 1996, 2010; Maier, 2008; Patterson, Greeson, & Campbell, 2009). Police continue to dismiss large numbers of sexual assault reports as unfounded and view many victims reporting sexual assault as unreliable witnesses of their own experiences (Crew, 2012; Doe, 2012; Doolittle, 2017). Rape crisis centres, which were once strong voices challenging institutions to reform their practices, have been significantly silenced by the weight of dwindling resources and the pressures to conform to criminal justice agendas and professionalized models of advocacy and counselling (Beres, Crow, & Gotell, 2009; Corrigan, 2013a; Doe, 2012; Russell, 2010). Accordingly, they have been pushed into the margins of the network of professionalized services for sexual assault victims.[1] As a result, medical and legal institutions continue to be sites where many sexual assault victims are re-victimized in their quest for medical treatment and justice.

This turn of historical events raises several questions: How did this network of professionals, expert practices, and forensic technologies for sexual assault response come to be? And why, in the face of all this change, do Clark and Lewis' groundbreaking findings still ring true? Answering these questions demands a historical trek through the controversies, debates, and political struggles in the past few decades over law, technoscience, victim advocacy, and sexual assault. It begs an analysis of changing practices and protocols, and new tools, technologies, and expertise around sexual assault forensics that arose during this time. This book takes up this task by examining the history of the Ontario Sexual Assault Evidence Kit (SAEK).

Forensics in a Box: The Ontario Sexual Assault Evidence Kit

Sexual assault evidence kits go by many names. Rape kits, crime kits, rape examination kits, and sexual assault investigation kits similarly refer to standardized kits that contain tools that medical examiners use to collect forensic evidence from sexual assault victims' bodies (Du Mont & White, 2007). Similar to kits in other jurisdictions, the Ontario SAEK includes swabs for collecting traces of foreign bodily fluids, vials for urine and blood samples, envelopes for hair and foreign debris, paper bags for clothing, and standardized instructions and forms for detailing injuries. A forensic exam guided by the SAEK involves multiple steps of swabbing, plucking, and combing, a process that can last up to 4–6 hours. If a sexual assault involved penetration of any kind, the exam can also include an internal examination, in which a medical examiner uses a variety of technologies to magnify and photograph injuries inside the victim's vagina, anus, or mouth (White & Du Mont, 2009). In the forensic exam, the victim's body is defined as the "crime scene," which is searched for evidence of sexual assault (Price, Gifford, & Summers, 2010). The forensic exam can be uncomfortable, painful, and re-traumatizing, as many reports from women who have experienced it have attested (Doe, 2012; Du Mont, White, & McGregor, 2009; Mulla, 2014).

The history of the Ontario SAEK provides a unique window into histories of controversy, activism, and medical, legal, and scientific practice around sexual assault. In Ontario, sexual assault forensics, victim advocacy, and treatment are notably guided and bound by provincial organizations, networks, and coalitions. Unlike other provinces in Canada, with the exception of Quebec, Ontario sexual assault kits are designed, distributed, and analysed by the provincial forensic laboratory, the Ontario Centre of Forensic Sciences (CFS).[2] Since the Ontario SAEK was first imagined in the late 1970s, new professional groups, sites for forensic evidence collection, scientific and medical practices, and technologies for forensic exams have developed around the SAEK. Sexual Assault Nurse Examiners (SANEs) in hospital-based sexual assault treatment centres, and some nurses and physicians in hospital emergency wards, now use the kit to collect bodily fluids and document physical injuries on victims' bodies. The treatment centres that SANEs work in are linked through a provincial network that was one of the first of its kind in Canada when it began in 1993. Outside of the hospital, community-based rape crisis centres in Ontario, which offer public education, counselling, and support for sexual assault victims, are connected through the Ontario Coalition of Rape Crisis Centres that began in 1977. These provincial organizations, networks, and coalitions played significant roles in the historical controversies around sexual assault forensics, advocacy, and treatment in the province, and to varying degrees have contributed to situating the SAEK as a central technology in medical and legal responses to sexual assault victims.

The SAEK is now commonly touted in law enforcement and forensic science circles as a tool that produces more reliable and credible forensic evidence of sexual assault. Its forensic contents, most particularly the DNA evidence in the bodily fluid samples, are praised for improving police investigations and criminal prosecutions. Police training manuals stress the importance and inherent trustworthiness of forensic evidence: "Physical evidence is that mute but eloquent manifestation of truth which rate high in our hierarchy of trustworthy evidence" (Baeza & Turvey, 2002, 169); and forensic science reports celebrate the powers of the DNA evidence inside the SAEK: "It cannot therefore be too strongly stated that evidence of this kind simplifies the administration of justice" (Breitkreuz, 2009, 1). Since the SAEK was first dubbed "Ontario's most successful rapist trap" (Crawford, 1984, A13) in 1984, the kit has secured a trusted place in medical and legal responses to sexual assault. Sexual assault police investigators and prosecutors often rely on kits to identify perpetrators of sexual assault and confirm the degree of force that a perpetrator used against a victim, the type of sexual activity that occurred, and the victim's inability to legally consent to the sexual encounter due to drugs and alcohol (Du Mont & White, 2007). Trust has seemingly developed around this technology to provide credible accounts of the traces left on victims' bodies. Unlike the victims who report sexual assault, the kit has seemingly gained credibility as a reliable "witness" of sexual assault, a technoscientific witness that can aid in the identification and conviction of sexual offenders.

Despite the apparent trust that the SAEK has secured, and the extensive network that has been built to support it, it is far from clear that the kit and the evidence it contains are actually improving victims' experiences in medical and legal institutions. Sexual assault forensic evidence in practice often has a negligible effect on outcomes of sexual assault investigations and trials (Du Mont, McGregor, Myhr, & Miller, 2000; Du Mont, Miller, & Myhr, 2003; Du Mont & White, 2007; McGregor, Le, Marion, & Wiebe, 1999), in contrast to popular claims to the contrary. Even further, the forensic evidence inside the SAEK can fuel defence lawyers' interrogation of victims' testimony in court and invite questions about a victim's medical and sexual history that are intended to be inadmissible in sexual assault trials (Burman, Jamieson, Nicholson, & Brooks, 2007; Rees, 2012; Temkin, 2000). The minimal legal benefit that forensic evidence often has for sexual assault prosecutions is in stark contrast to the trauma, pain, and discomfort that many victims experience during the forensic exam (Corrigan, 2013b; Doe, 2012; Mulla, 2014; Rees, 2012, 2015). In the face of this evidence, how did the SAEK gain and maintain a central place in medical and legal practice as a trusted tool for forensic evidence collection? How did it become defined in medical and legal practice as "the right tool for the job" (Casper & Clarke, 1998, 258) for collecting evidence of sexual assault and identifying and convicting its perpetrators? Assembling the SAEK as the apparent "right tool" for sexual assault investigation and prosecution involved assembling a network of actors that victims now face when they report their sexual assault to police. This book traces the history of the SAEK and its assembly as the trusted technoscientific witness of sexual assault.

Law, Science, and Anti-Rape Activism

A historical look at the SAEK opens up questions about law, science, technology, medicine, and anti-rape social movements, and the complicated relations between them. The intersections between science and law have been the focus of a wealth of literature in science and technology studies and socio-legal studies.[3] Likewise, the complex relationships between science, medicine, and social movements have been the subject of much work in science studies and social movement research.[4] These literatures provide some useful frames for understanding how practices in science, law, and medicine converge, and the role that technology and social movements can play in those convergences.

Science and law, according to Lynch, Cole, McNally, and Jordan (2008), are increasingly being drawn together as more scientific and technical experts are being pulled into legal disputes and as new sciences and technologies are challenging legal distinctions and definitions. When science meets law, they argue, rich insight can be gained about how these two seemingly disparate worlds work together. Law and science, according to Jasanoff (1995), are commonly misunderstood as separate and distinct worlds. Science pursues truth, whereas law pursues justice, so the conventional wisdom goes. Jasanoff suggests that there are indeed notable differences between how scientists and legal practitioners develop facts: while both law and science aim to get facts right, law's fact-finding is distinctly motivated by ideals of fairness and efficiency. However, these differences between law and science, in Jasanoff's view, are often overstated. Assuming a fundamental "cultural clash" (7) between law and science masks the ways that cultures of law and science are often "mutually constitutive" (8). Instead, she argues, law and science can be more fruitfully analysed as institutions that work together to produce knowledge about society.

Similar arguments have been made about the analytical benefits of tracing the relations between science, medicine, and social movements. Epstein (1996) argues that such an analysis can reveal how some social movements have been shaped by their encounters with medical science and, likewise, how political pressures from social movements have influenced medical science. The particular interconnections between feminist social movements, medical science, and technology come to life in Murphy's (2012) historical work on the women's health movement of the 1970s and 1980s. She charts how the movement turned to technoscience to gain control over women's reproduction and redefine women's medical care, and the tense and productive relations that emerged between feminism and technoscience as a result. Following similar lines of inquiry, this book looks into the SAEK's past to see how law, medicine, and science have historically coalesced to produce legal "facts" about sexual assault, and the complicated role that feminist anti-rape movements played in politicizing, maintaining, and denouncing these forensic technoscientific practices.

Feminist anti-rape movements have had a long and complex relationship with law, criminal justice, and the state. Corrigan (2013a) and Bumiller (2008) have similarly traced anti-rape movements in the United States and how they formed what Corrigan calls, "a prickly political alliance" (38) with law and the state. By relying heavily on legal reform and crime control as tools for social change, Corrigan argues that the anti-rape movement fell out of step with other left-leaning social movements and into step with conservative criminal justice agendas. As a result of its alliance with law and the state, the movement was radically transformed. The movement, according to Bumiller, was co-opted to align with neoliberal state interests and, according to Corrigan, demobilized by its alliance with the priorities of law and criminal justice.

In Canada, activists in the anti-rape movement had a particularly complicated relationship with law and criminal justice. While some activists in the 1970s and 1980s worked with actors in the criminal justice system to develop reforms in institutional practices, protocols, and law, others actively resisted these alliances. Heated debates waged within the movement over the efficacy of working with institutions of law enforcement and medicine. Hospitals, law enforcement organizations, and government agencies did not always welcome activists, and as a result, activists often had to struggle to be included in discussions about institutional reforms and for the right to work as volunteer victim advocates in hospitals, police stations, and courtrooms. As pressures to professionalize victim services grew in the 1990s and 2000s, volunteer advocates in rape crisis centres were increasingly pushed out of medical and legal spaces. The Canadian anti-rape movement's alliance with law and criminal justice has thus been a site of contentious negotiations and struggle.

The story of the SAEK provides a window into these complex relations. The contemporary kit draws together the seemingly independent worlds of law, science, and medicine. As an object that is used in hospitals, stored in police stations, analysed in forensic labs, and employed in courtroom trials, the kit pulls together practices in medicine, science, and law to build legal truths about sexual assault and its perpetrators. With its standardized components, the kit facilitates communication and coordinated action between medical examiners, forensic scientists, and police investigators. The work in hospital exam rooms, forensic labs, police stations, and courtrooms becomes part of a chain of action in which the forensic contents of the kit are assembled, analysed, interpreted, and interrogated. To use Star and Griesemer's (1989) term, the kit is a "boundary object" (393) that facilitates this coordinated action.[5]

Star and Griesemer (1989) describe boundary objects as scientific objects that occupy multiple intersecting worlds and enable coordination between them. They are "objects which are both plastic enough to adapt to local needs and the constraints of the several parties employing them, yet robust enough to maintain a common identity across sites ... They have different meanings in different social worlds but their structure is common enough to more than one world to make them recognizable" (393). As the kit moves through medical, scientific, and legal spaces, it maintains a common identity as a standardized tool that is seemingly necessary for forensic evidence collection in sexual assault cases. While this common identity coordinates how nurses, police, forensic scientists, and lawyers work with the kit, in medical, scientific, and legal sites the kit often takes on different meanings for different actors. In the eyes of some medical examiners, the kit is a tool that simplifies evidence collection, whereas for many forensic scientists it is a tool for assembling reliable samples, and for most police investigators it is a tool for producing necessary corroborative evidence.

For some anti-rape activists and advocates, the kit is a symbol of progress and institutional reform that has potential to benefit some victims; for others, it is an extension of patriarchal and racist institutions that systematically fail to meet victims' needs. The kit is thus a complicated boundary object that not only coordinates action, but also simultaneously embodies these many tensions between the worlds of medicine, science, law, and anti-rape advocacy.

Thinking about the SAEK as a *technoscientific* witness draws attention to the complexities of this boundary object. The term usefully points to how the kit blurs the lines between science and technology and works in different ways with different meanings in medical and legal spaces. The term technoscience, unlike science and technology, implies an unsettled state, in which definition and meaning are still being contested and negotiated (Latour, 1987). Thinking about the kit as a technoscientific witness opens up the possibility of seeing the controversies about the kit's meaning and purpose, and more specifically, how the meaning of the kit has been contested and negotiated through decades of activism, institutional reform, and changes in medical, scientific, and legal practice in Ontario.

Mulla (2014) proposes that sexual assault forensic exams have created a new domain that is not medical or legal but is, instead, distinctly *forensic*. Others call this new domain a *medicolegal* one (Parnis & Du Mont, 2002; White & Du Mont, 2009). Through the SAEK's history, this book takes up unanswered questions about how this domain came to be and what controversies and political tensions were involved in its creation. The SAEK's history provides an opportunity to examine how science, law, medicine, and feminist social movements became intertwined in the quest to identify and convict perpetrators of sexual assault. This history bears not only the troubled relations between the anti-rape movement and law and medicine, but also anti-rape activists' complicated relationships with forensic science and technology. How did technoscience become wrapped up in anti-rape activists' struggles to improve medical and legal responses to sexual assault? What tensions were embedded in their uneasy relationships with technoscience and the law? And how did these relationships shape the movement and the work of anti-rape activists? A historical look at the SAEK casts light on how law, medicine, and science met in the effort to develop a credible forensic technology for sexual assault prosecutions, and how anti-rape social movements became caught up in those struggles.

The Politics of the SAEK

The SAEK is not a neutral object. Instead, it is a technology that reflects and maintains the histories and politics of the worlds of which it is a part. Legal, scientific, and medical forms of knowledge have long been associated with the virtues of objectivity and neutrality. However, as many feminist scholars have argued, law, science, and medicine are similarly rooted in ways of knowing that reflect and reproduce white, male privilege and power (Haraway, 1997; Harding, 1991, 2008; Keller, 1985; MacKinnon, 2005; Smart, 1989). The history of sexual assault law and medicine illustrates this clearly. Sexual assault victims, most particularly non-white, female sexual assault victims, have historically had little credibility in medical exam rooms and courtrooms. In law, forensic medicine and science have been viewed as necessary counterbalances to women's presumed untrustworthiness and lack of credibility (Backhouse, 2008; Dubinsky, 1993). The long history of legal actors placing faith in forensic science and medicine to reveal the truths about sexual assault, and the continued doubt of sexual assault victims that fuelled it, are politics that are now embedded in the contemporary SAEK.

Langdon Winner's (1980) widely cited article, "Do Artefacts Have Politics?," makes a strong case for viewing technologies as political artefacts. Technologies are "ways of building order in the world" (127), he argues, which establish, reflect, and reinforce forms of power and authority. He writes: "If our moral and political language for evaluating technology includes only categories having to do with tools and uses, if it does not include attention to the meaning of the designs and arrangements of our artifacts, then we will be blinded to much that is intellectually and practically crucial" (125). The material design of technologies, and the consequences that these designs have on social relations, do matter. Being attentive to the politics of the SAEK calls for a historical look into the SAEK's changing material designs, and the role that the kit has played in reflecting and reinforcing a politics of distrust of sexual assault victims in the criminal justice system.

A growing number of scholars in Canada and beyond have begun interrogating technologies and practices in sexual assault forensic exams, the values embedded within them, and their consequences for victims. In 1997, Georgina Feldberg broke significant empirical ground with her sharp critique of the Ontario sexual assault forensic exam (Feldberg, 1997). The exam, she argued, represents a misguided impetus to construct sexual assault in seemingly objective scientific and technical terms. It produces evidence that has little relevance in court and ignores the more significant pyscho-social harms of rape, she contended. Building on this work, Canadian scholars Janice Du Mont and Deborah White[6] have since argued that sexual assault kits rest on outmoded legal requirements of corroborative evidence and perpetuate outdated notions of sexual assault as a physically violent act perpetrated by strangers (Parnis & Du Mont, 1999, 2006). Through their insightful qualitative and quantitative research, they have investigated the uses and efficacy of forensic exams, exposed their impacts on victims of sexual assault, and examined the effectiveness of sexual assault nursing programs, discretionary and standardized practices in forensic exam rooms and laboratories, and victims' varied experiences of SAEK exams (Du Mont & Parnis, 2001, 2003; Parnis & Du Mont, 1999, 2002, 2006; White & Du Mont, 2009). Taking up some of these themes, Jane Doe (2012) has interrogated the harms of Canadian sexual assault kit exams through the perspectives of advocates and women who have undergone the forensic exam, and found that the kit rarely benefits women who have experienced sexual assault. Scholars outside of Canada have similarly called into question the values and consequences of sexual assault forensic exams. Most particularly, Corrigan (2013b), Rees (2012, 2015), and Mulla (2014) have drawn attention to the guise of scientific objectivity that sexual assault forensic exams operate under and its negative impacts on sexual assault victims in hospitals and criminal trials. This existing literature paints a picture of contemporary forensic exams that rarely live up to their promise of serving sexual assault victims' interests. It opens the door to historical questions about the political histories embedded in the contemporary forensic exam and the role the SAEK has played in building and maintaining medicolegal networks in which victims have little credibility.

The view of the SAEK in medical and legal practice as an objective and neutral technology obscures its political origins in patriarchal and racist legal histories of doubting and dismissing women's reports of rape. It hides the politics of its design and continued use. Most importantly, it obscures the SAEK's work in a medicolegal system that continues to harm sexual assault victims and fail to meet their needs. Shedding light on the SAEK's origins reveals its politics and how it gained its contemporary role as a technoscientific witness of rape in medical and legal practice.

A Technoscientific Witness

Law enforcement and forensic science communities have spent more than a century searching for a reliable way to determine the identity of perpetrators of crime (Aronson, 2007). Early-twentieth-century methods of fingerprinting and anthropometry[7] involved analysing perpetrators' bodily characteristics in an attempt to identify them and track their crimes. For much of the twentieth century, fingerprinting and anthropometry were considered the best criminal identification methods that forensic science had to offer (Cole, 2001). These techniques fell out of favour, however, when scientists began disputing their reliability and accuracy and when newer, seemingly more credible, technologies for criminal identification emerged. Now decades later, forensic DNA analysis is considered to be the gold standard for identifying perpetrators of crime, despite the heated debates about its reliability when it was first introduced to the criminal justice system in the late 1980s (Aronson, 2007; Gerlach, 2004; Lynch, Cole, McNally, & Jordan, 2008). Investigators and prosecutors now rely on genetic material in biological fluids and skin cells that a perpetrator leaves at a crime scene for forensic identification. Forensic DNA analysis, like the forensic methods before it, rests on the understanding that a perpetrator's bodily characteristics, whether the lines on their fingertips or their genetics, can reveal their identity (Cole, 2001).

The physical evidence that perpetrators leave behind is often described in forensic science circles as "a silent witness" of the crime (Bieber, 2004; Fourney, as cited in National DNA Databank, 2002). While the silent witness is now a popular moniker for DNA evidence, this phrasing was in existence decades ago. In 1953, Paul Kirk, author of *Criminal Investigation: Physical Evidence and the Police Laboratory*, described the powers of physical evidence to reliably witness criminal activity:

> Wherever he steps, wherever he touches, whatever he leaves, even without consciousness, will serve as a silent witness against him. Not only his fingerprints or his footprints, but his hair, the fibers from his clothes, the glass he breaks, the tool mark he leaves, the paint he scratches, the blood or semen he deposits or collects. All of these and more, bear *mute witness* against him. This is evidence that does not forget. It is not confused by the excitement of the moment. It is not absent because human witnesses are. It is factual evidence. Physical evidence cannot be wrong, it cannot perjure

itself, it cannot be wholly absent. Only human failure to find it, study it and understand it can diminish its value. (Paul Kirk, as cited in Mulla, 2014, 37, emphasis added).

Physical evidence, so this narrative goes, cannot embellish, lie, or forgot like the humans around it. Rather, it reliably "witnesses" the crime and gives factual "testimony" of criminal behaviour.

The SAEK, much like the physical evidence it collects, acts as a witness. Instead of witnessing the crime as the physical evidence seemingly does, the kit works alongside physicians and nurses to *witness* the traces a perpetrator leaves on a victim's body – the so-called scene of the crime. Its contents retain a memory of the injuries and bodily fluids that a perpetrator left behind on the victim's body and clothes. After being used to collect evidence in the medical exam room, the kit travels through the hands of police investigators, forensic scientists, technicians, and lawyers who analyse and interrogate its contents. If the kit reaches the courtroom, it then acts as a witness of the traces of rape observed on the victim's body in the forensic exam room. Without the medical examiner or victim having to speak, the kit's contents provide a visual testimony of the traces of rape left on the victim's body.

The kit provides different kinds of visual depictions of rape: a victim's injuries are made visible with drawings and photographs, and the perpetrator's identity becomes visible with a DNA profile and statistical probabilities. While the physical evidence of injuries can be easily "seen" by lawyers and judges in the courtroom, the DNA evidence in the kit becomes visible through other means. Jasanoff's (1998) detailed analysis of the O.J. Simpson trial reveals how DNA evidence is made visible in the courtroom with scientific instruments and practices. She argues that scientists and non-scientists in court do not always "see" DNA evidence in the same way, and that non-scientists in the courtroom have to be trained to "see" and understand the DNA evidence in the way that scientists do. It is through forensic scientists' expert testimony and their statistical calculations that DNA evidence and a perpetrator's DNA profile become visible in the courtroom. Jasanoff argues that in court, visualization and credibility go hand in hand. To use her words, visualization is one of the ways in which "scientific evidence achieves credibility – and so gains, for the purpose of legal decision making, the status of fact" (716). The kit's capacity to provide visual depictions of rape is thus at the core of its presumed credibility in sexual assault investigations and trials. Although the majority of sexual assault cases do not end up in the courtroom and are dismissed by police long before as unfounded (Crew, 2012; Doolittle, 2017; Johnson, 2012), the anticipated destination of the courtroom for kit evidence gives the kit its meaning and presumed value (Mulla, 2014).

Unlike victims and other human witnesses, the kit is trusted not to forget, never to lie, and to always provide the objective facts of sexual assault. The kit's credibility as a reliable technoscientific witness of sexual assault developed alongside ongoing legal practices of distrusting victims who report sexual assault and labelling them as unreliable witnesses of the crimes committed against them. The credibility of one witness, the technoscientific witness, was thus built on the lack of perceived credibility of another, the victim.

The contemporary SAEK's credibility in medical and legal practice is tied to its presumed ability to witness. Scientific witnessing, according to Haraway (1997), first became associated with the virtues of objectivity and credibility during the birth of experimental science in the seventeenth century. Building on Shapin and Schaffer's (1985) historical theorizing on early experimental science, Haraway describes how scientific practice became associated with a form of "modest witness[ing]" (32), in which scientists made their bodies and selves invisible within their accounts of the natural world. Self-invisibility was a form of modesty that defined scientists as legitimate, credible witnesses who could assert believable, scientific "matters of fact" (29) about the natural world. Technologies played a role in the construction of the scientist as a credible, modest witness, according to Shapin and Schaffer. Technologies were used to widen the apparent distance between a scientist and their account of the natural world; with technologies, Shapin and Schaffer write, a scientist could say, "It is not I who say this; it is the machine!" (25). Haraway (1997) argues that the modest witness of early experimental science was a gendered, classed, and raced witness. The form of modesty required for scientific witnessing in seventeenth-century England, was a virtue to which only white men of the upper classes could aspire and achieve. Women, working-class men, and men of colour were believed to be unable to achieve the modesty necessary for objective scientific witnessing. Women in particular were seen as unable to transcend their embodied beings as male scientists could. The legacy of this history, Haraway contends, is a modern conception of science that has at it core a masculine version of scientific modesty that is premised on disembodied and self-invisible truth telling.

Taking this understanding of scientific witnessing into the realm of the SAEK points to the origins of its credibility as a technoscientific witness of rape. The SAEK's credibility rests on its presumed capacity to offer disembodied, self-invisible accounts of the traces of rape on the victim's body. In the courtroom, the kit is the quintessential *modest witness* of rape. Its apparent distance from human hands and minds gives its contents the credibility to assert legal matters of fact about sexual assault. Alongside physicians and nurses in the exam room, the SAEK participates in a form of witnessing of a victim's body, in which the victim has little credibility and her body is treated as a crime scene to be combed for evidence. It is a form of witnessing that reinforces and re-enacts gendered, raced, and classed histories of sexual assault law and medicine.

To suggest that the kit participates in *witnessing* traces of sexual assault on victims' bodies implies that the kit has agency, or at least enough agency to be deserving of the verb. I propose that the kit is not merely a passive object that is simply used by human actors. Rather, it works alongside nurses, doctors, police, scientists, lawyers, and other tools and technologies to make forensic evidence of sexual assault. While the SAEK is literally put to work by human hands, it simultaneously exerts influence on what those human hands do in the course of making forensic evidence. It shapes and constrains how evidence is collected and analysed, and the meaning that the evidence has in investigations and courtrooms. It enacts victims' bodies as crime scenes and enforces expectations on victims to comply with forensic investigative procedures.

To think about the SAEK as an actor, I draw on traditions in science and technology studies that have sought to understand the materiality of technoscientific practice by exploring the roles that both humans and non-humans play in this work. Latour (2005), Callon (1999), Law (2004), and other scholars in actor-network theory have described technoscientific work as inherently heterogeneous, involving the action of both human and non-human actors. Ignoring how non-humans act with and alongside human actors in scientific and technical practice, they argue, is to miss key elements of stories about technoscience. Callon and Law (1997) describe these stakes further:

> Often in practice we bracket off non-human materials, assuming they have a status, which differs from that of a human. So materials become resources or constraints; they are said to be passive; to be active only when

they are mobilized by flesh and blood actors. But if the social is really materially heterogeneous then this asymmetry doesn't work very well. Yes, there are differences between conversations, texts, techniques, and bodies. Of course. But why should we start out by assuming that some of these have no active role to play in social dynamics? (168)

Their assertion that non-human objects have the potential to act in technoscientific practice is undoubtedly a provocative one. Law, Callon, Latour, and others in the actor-network theory tradition have been criticized for taking an overly symmetrical view of humans and non-humans, which does not do enough to appreciate the differences between them (Haraway, 1997; Idhe, 2002). However, their work usefully raises a call to acknowledge the role that non-humans can play in technoscientific and criminal justice practice (Moore & Singh, 2015; Robert & Dufresne, 2015a).[8] Recognizing the potential for non-humans to act does not *have* to go hand in hand with ignoring the differences between humans and non-humans. It is possible to recognize the role that non-human actors play in technoscientific practice without overshadowing the responsibility and agency of human actors. Understanding how human and non-human actors work alongside one another usefully draws attention to the materiality of technoscientific work, and the material relations between humans and non-humans.

Sexual assault forensics demands an attention to the material. Bodies, fluids, swabs, speculums, DNA profiles, microscopes, *and* the SAEK are present in forensic practice and can shape the forensic evidence that is produced. Although it is human actors who have the responsibility to act and the agency to change the course of that action, these material objects and entities can and do act in forensic practice. Thinking about the kit as an actor in relations with many other human and non-human actors makes its work in medical and legal institutions more clear. It is no longer just an innocent, neutral object. Instead, it is a material object that acts with others in a system of practices that rarely works in victims' interests.

The kit's capacity to act is not inherent. Nor does the kit act alone. It is instead part of a medicolegal *network* of actors who collectively respond to victims' reports of sexual assault. The metaphor of a network can be useful for understanding the ways in which social and technical worlds are co-constructed (Latour, 1987, 2005; Halfon, 1998). Thinking about networks of social and technical relations reveals how technologies are not simply material objects, but are instead "social processes through which many disparate social, material, and rhetorical elements are brought together in mutually constitutive ways" (Halfon, 1998, 28). Networks can be thought of as heterogeneous "webs of relations" (Law, 2007, 1) between human and non-human actors. Latour (2005) warns that the term network should not be taken to imply an entity that exists outside of practice. The term *network* does not specify a particular shape (Latour, 2005), nor does it imply a functional one (Law, 2003). Rather, it is a tool for description (Latour, 2005). Here, I use the term as a tool for understanding the SAEK as a boundary object that brings together the worlds of science, law, medicine, and advocacy, and the practices and material relations between actors within these worlds.

When hospital staff, police investigators, scientists, and lawyers work with the SAEK and other tools and technologies to assemble forensic evidence of sexual assault, they form a network of practices, which victims become intertwined in when they report a sexual assault and consent to a SAEK exam. Within this network, there are complicated power relations between advocates, victims, and medical and legal professionals, and ongoing controversies about sexual assault treatment, advocacy, law, and forensics. Since the kit was first introduced to Ontario, relations in this network have significantly changed with the pressures of these controversies and the development of new experts, technologies, and scientific and medical practices around the kit. New laboratory techniques introduced in the 1980s transformed how the kit was analysed and new experts pushed advocates in rape crisis centres further into the margins of medical and legal practices. Victims navigate a very different network today than they did forty years ago.

The kit is what it is *because of* the network of relations that it is a part of. The kit has meaning as a technoscientific witness because it is part of a network of actors and practices that make it so. The technoscientific witness is thus not reducible to the cardboard box, swabs, and vials that make up the kit's material form. Rather, it exists and has meaning and agency because of its relations with others. Actors do not exist alone. Instead, they exist relationally, for as Mol (2002) says, "to be is to be related" (54). My objective in this book is not simply to observe the kit through history, but rather to see how the kit acts and is enacted in practices with other actors.

Although I understand the SAEK to be an actor, I do not do so to release human actors from their responsibility for the ethical treatment and care of sexual assault victims. While there are analytical benefits to viewing the kit as an actor working alongside other human and non-human actors, Haraway (1997) cautions that there is a danger in drawing attention to non-human actors. She writes that although both human and non-human actors have agency, "our relationality is not of the same kind of being. It is people who have the emotional, ethical, political, and cognitive responsibility inside these worlds. But nonhumans are active, not passive, resources or products" (Haraway, 2000, 10). By positioning the SAEK as an actor that has been brought into being through decades of medical and legal practice, this study demands accountability and responsibility from human actors for the SAEK and the practices in which it is involved. Seeing the kit and its network as entities that emerged out of material practices makes it clear that its current form was never inevitable. In fact, it "might have been otherwise" (Hughes, 1971, 552), to use Hughes's commonly cited phrase, if law enforcement had not been shaped by conceptions of victims as inherently unreliable witnesses or if work in laboratories, hospital exam rooms, and courtrooms had been organized in ways that more adequately served victims' interests. If the SAEK *could have been* otherwise, then it follows that it *can be* otherwise. Trekking through the kit's past makes clear not only the pressing need to build better futures for sexual assault victims in and outside of the criminal justice system, but also the many obstacles of doing so.

Diffracting the SAEK

Histories of technoscience matter. Donna Haraway (1994) argues that "if technology, like language, is a form of life, we cannot afford neutrality about its constitution and sustenance" (62). How technologies are built and sustained in practice reveals much about the social, political, economic, and cultural worlds of which they are a part (Bijker, 1997; Law, 2002; Winner, 1986). Many scholars have charted histories of technological objects within science and technology studies (Bijker, 1997; Casper & Clarke, 1998; Dugdale, 2000; Takeshita, 2012; Winner, 1986). This work has shown that technologies do not materialize from single discoveries of inventors working in isolation, as popular notions of technological development might suggest. Instead, technologies emerge out of technical, scientific, social, political, economic, and cultural processes and bear the traces of those contexts. The SAEK's history provides an entry into the many contexts from which it was born. As will become evident, the kit's origin story is not reducible to one person or one event. Rather, the kit was assembled and reassembled through controversies about medical treatment of sexual assault victims, police handling of sexual assault reports, sexual assault laws, and scientific techniques and tools in forensic laboratories. The network of practices of which the kit is a part now wields great power over victims who report their assaults, and therefore we cannot afford to be neutral about how the SAEK came to be or what it currently is.

To guide this trek through the SAEK's history, I employ Donna Haraway's (1997, 2000) diffraction metaphor. The task of feminist technoscience studies, Haraway contends, is not simply to reflect dominant narratives of technoscience, but instead to *diffract* technoscience so that it becomes possible to see what it is made of and to imagine more ethical and responsible alternatives. Diffraction involves "see[ing] both the history of how something came to 'be' as well as what it simultaneously is" (Goodeve as quoted in Haraway, 2000, 108). Diffraction is a metaphor for an act of interference, as Takeshita (2012) has explained, that is akin to breaking up beams of light to reveal the many coloured rays within them. Diffraction, for Haraway and others who have taken up her work, involves fragmenting single visions of technoscientific objects to reveal what has been obscured or made invisible in their histories and contemporary uses.

The single beam of light that this study diffracts is the popular claim that the contemporary SAEK is a useful and necessary tool for identifying and convicting perpetrators of sexual assault. This study breaks up this beam of light to reveal all the coloured rays of light it contains: the many political histories of struggle, controversy, uncertainty, and shifting medicolegal practices that are embedded in the SAEK. By diffracting the SAEK, it becomes possible to see how the kit came to be, and all the many things that make it what it currently is. More specifically, we can see how the kit gained status as a technoscientific witness that often wields more credibility than the victims who report sexual assault. Diffracting the SAEK in this way requires trudging through rough terrains of practice and controversy, much like, as Latour (2005) observes, an ant does over and through hills of dirt and grass. Uncertainties, tensions, and debates in law, feminism, and forensic science become visible through this diffracted visioning of the kit, which lays the necessary ground for imagining more ethical and alternative ways of organizing medicolegal practice around sexual assault.

While Haraway (2000) makes clear that the aim of diffraction is to reveal the many coloured rays of technoscience, not all rays of light can be equally as bright. Law (2003) asserts that "any way of imagining technologies is partial" (2). Any attempt to sketch the SAEK's history will therefore be imperfect, and necessarily so. Fitting the SAEK's story into the pages of this book required strategic choices about what parts of the history would form the foreground of the book's narrative and what parts would be pushed into the background. Although some victims' narratives of the SAEK exam are featured here, I do not focus specifically on victims' experiences or perceptions of the kit. Other scholars have taken up this important task (Doe, 2012; Mulla, 2014; Du Mont, White, & McGregor, 2009). My focus is instead on the network of actors responding to victims' reports of sexual assault: advocates, doctors, nurses, police, lawyers, and forensic scientists, and the many tools, texts, and technologies that they work with. I am interested in what these actors have done and currently do when a victim reports a sexual assault, and how the SAEK was and is implicated in this work.

Methodology

Much of the SAEK's history has been forgotten or buried in archives and the memories of people who were part of its development. High turnover rates in forensic nursing and sexual assault policing[9] have left few people in medical and legal fields whose memories of sexual assault forensics reach back beyond recent years. In addition, historical records on the kit's early years are difficult to find; many are scattered across different archives, are protected by privacy regulations, or are in personal collections that are not publicly accessible. The kit's contemporary contexts are similarly difficult to trace. Much of the forensic work in sexual assault cases goes on behind closed doors of medical exam rooms and forensic laboratories.[10] To sketch the SAEK's past and present, I had to uncover and weave together many threads of historical and contemporary practice. This historical "muckracking" (Marx, 1972, 1) through the institutions within which the kit operates called for a wide range of data.

I conducted sixty-two interviews with retired and currently employed sexual assault nurses, forensic scientists, police investigators and administrators, Crown, defence, and civil lawyers, victim services staff, and advocates in rape crisis centres in Ontario (for a detailed breakdown of participants see the appendix). The participants were located in twenty-six cities and communities across the province. Because most specialized sexual assault services in hospitals, police organizations, and community-based organizations in Ontario are located in urban centres, the interview participants were all situated in urban communities, ranging in population size from under 20,000 to over two million. The vast majority of participants at the time of being interviewed were currently employed in law enforcement, forensics, and/or the sexual assault services sector, while the remaining few had retired from long-term careers in their fields. Participants' professional experience in their respective fields ranged from under a year to over forty years (for a summary see the appendix). Although my primary focus was on the actors who respond to reports of sexual assault, three of the advocates in rape crisis centres self-identified as sexual assault survivors and shared their experiences of the forensic exam, two of which occurred in the 1990s and the other, in the late 1970s. The sixty-two interviews varied in length from 30 minutes to over 2½ hours, and all interviews were audio-recorded and transcribed verbatim, with the exception of four interviews that were not audio-recorded at the request of the participants. Interview questions focused on historical and contemporary sexual assault law, advocacy, forensics, and treatment, particularly as they related to the SAEK. The majority of participants are quoted in the pages of this book. To preserve confidentiality, I have not included any references to the specific communities or organizations where the participants work.

In addition to the interview data, I also collected archival records from six Ontario-based archives that spanned national, provincial, municipal, and institutional levels. Reports, manuals, pamphlets, posters, letters, press releases, newspaper clippings, audio recordings, and training videos formed an extensive sample of historical materials. I also collected contemporary texts, including legal case files, media articles, and national, provincial, and municipal government reports. I used Access to Information requests to gain access to historical archival and contemporary texts that are protected by privacy regulations. To complement this array of textual and interview data, I conducted tours of a sexual assault treatment centre, forensic laboratory, and a police investigation unit. These tours brought to life the tools, technologies, and physical spaces that participants and the texts described.

My aim of diffracting the SAEK called for a unique analytical approach to the empirical data. Law (2004) argues that many existing social science methods for data analysis do a poor job of capturing the "messiness" (p. 5) of technoscientific practice. These methods, he suggests, often seek to reduce rather than expose the complexity, irregularity, and disorder within social-technical worlds. Methods that reduce reality to a finite number of themes or commonalities have limited use for diffraction (ibid.). Diffraction instead calls for methods of data analysis that make complexities visible (Haraway, 1997). I analysed the data in this study in ways that sought to reveal the SAEK's diffracted histories. I examined the interview and textual data with a particular eye to historical moments when the story shifted, controversies erupted, uncertainties waged, and practices changed. Through this approach, I assembled a story about the historical complexity embedded within the contemporary SAEK.

In this history of the SAEK, I have chosen to use the term *victim* to describe those who have experienced sexual assault. There is a long history of debate among feminist scholars about the politics of language around sexual assault. Although victim was a term commonly used by feminist activists in the 1970s (Brownmiller, 1975; Clark & Lewis, 1977; Medea & Thompson, 1975), it was criticized in the late 1980s and 1990s for falsely denoting passivity and denying women agency (Allard, 1997; Kelly, 1988; Wolf, 1993). Many scholars, activists, and advocates have more recently adopted the term survivor and have argued that it more accurately captures women's agency and strength during and after a sexual assault (Ontario Coalition of Rape Crisis Centres, 2015; Ullman, 2010). Both terms, however, have been criticized for limiting women's narrative agency and for implying a passive female body that is victimized by or survives an active male body (Doe, 2012; Spry, 1995). It is clear that neither term captures the breadth of women's experiences of sexual assault. There is a pressing need for new, more inclusive language. However, in this book, I have opted to use the imperfect term victim to reflect the historical and contemporary contexts in which the kit operates. In the 1970s, when the kit was first being imagined and designed, victim was a widely used term in rape crisis centres and medical and legal institutions. Forty years later, victim continues to be a well-used term in law enforcement. In the system of practices in which the kit operates, those who have experienced sexual assault are often defined as victims, and, as this study will show, are routinely denied the agency that the term survivor might suggest. As this book's analytic gaze falls on these institutional practices, it is fitting to employ the term victim.

At various points throughout this text, I use a female pronoun to refer to victims of sexual assault. I do not do so to discount that men and genderqueer individuals can be and are sexual assault victims. However, Canadian victimization surveys and crime statistics illustrate that women experience substantially higher rates of sexual assault: one estimate is that women are 5.6 times more likely to be sexually assaulted than men (Brennan & Taylor-Butts, 2008). Women are also more likely to report their experiences of sexual assault to hospitals and police (Light & Monk-Turner, 2009; Washington, 1999). My language usage reflects these trends.

Organization of the Book

This book is loosely organized around decades of the SAEK's life in Ontario and the heterogeneous domains of action and controversy associated with the SAEK that characterized each decade. Beginning in the 1970s, the pages of this book sketch the controversies around sexual assault and the SAEK that started on the streets, in rape crisis centres, and government boardrooms in the 1970s, and moved to hospital emergency wards in the 1980s and 1990s, and to scientific laboratories, police stations, and courtrooms in the 1990s and 2000s. By following the action around the SAEK through time and space, I show how the kit and its contemporary network were built and how the tool gained its prominent place in contemporary medical and legal practice. While the chapters of this book are organized to roughly reflect these seemingly disparate domains, each chapter features a wide range of different actors – activists, hospital staff, police investigators, government officials, and scientists – to show the interconnections between the worlds of law, science, medicine, and anti-rape advocacy. In doing so, I illustrate how the SAEK has, throughout its history, acted as a boundary object bringing seemingly disparate domains of action together. Together, these chapters reveal how the SAEK was assembled and reassembled into a technoscientific witness of sexual assault.

Chapter 2 opens in the early 1970s, the beginning of a decade of rising anti-rape activism, heated controversies over medicolegal practice, changing forensic technologies and sciences, and firmly entrenched legal practices of labelling victims – particularly women – reporting sexual assault as unreliable and non-credible witnesses. This chapter traces how these contexts inspired the design of the first Ontario SAEK. I examine anti-rape advocates' work in rape crisis centres, hospitals, and government consultations, which laid the foundations for the SAEK. I use Akrich's (1992) notion of inscription to describe the SAEK's design as a process that involved *inscribing* meanings, practices, and histories into the first kit. The chapter shows how this design work involved medicolegal professionals and anti-rape activists imagining a technoscientific witness that promised to give more credible witness testimonies of sexual assault.

The efforts in hospitals, courtrooms, and rape crisis centres in the 1980s to stabilize the first SAEK in the Canadian criminal justice system is the subject of chapter 3. Set against the backdrop of the 1980s restructuring of medical institutions and professionalization of rape crisis centre advocacy, this chapter follows the controversies that the SAEK's arrival in Ontario sparked among nurses, doctors, police, lawyers, and rape crisis centre workers. Through these controversies, the chapter examines the development of new expertise, experts, and expert spaces for sexual assault care that changed medicolegal practice and rape crisis centre advocacy. I illustrate how through the webs of controversy, relations in the SAEK's network seemingly stabilized and the kit gained stability as a credible technoscientific witness of sexual assault in the Canadian criminal justice system.

Chapter 4 begins in the late 1980s amidst dramatic changes in forensic identification practices in North America. This chapter examines the turbulent rise of forensic DNA typing in Canada and reveals how the Sexual Assault Evidence Kit was *re*assembled alongside DNA typing into a tool for identifying perpetrators of sexual assault. I sketch the controversies related to forensic identification in sexual assault cases in the late 1980s and 1990s and the corresponding ongoing legal battles over sexual assault victims' trustworthiness. I follow anti-rape activists' stark opposition to DNA analysis, and the legal and scientific debates about the reliability of forensic DNA analysis. This chapter argues that with the development of DNA analysis, the SAEK was transformed into a *genetic* technoscientific witness that, in the eyes of many police investigators, lawyers, judges, and policymakers, had more power to give objective and reliable witness testimony of sexual assault than its predecessor and the victims whom many hoped it would protect.

Turning to the contemporary SAEK, chapter 5 addresses the questions: how does the SAEK act as a technoscientific witness in contemporary sexual assault investigations and prosecutions, what are the costs of its stability as a technoscientific witness, and to whom? This chapter traces the changes in sexual assault nursing, advocacy, and expertise and describes the medical and legal practices in which the contemporary SAEK acts as a technoscientific witness of sexual assault. I describe how police and defence lawyers routinely rely on the SAEK as an objective witness to test the veracity of victims' reports of sexual assault.

Building on existing literature on the SAEK, this chapter reveals how the technoscientific witness rarely benefits victims of sexual assault and can in some cases become a tool for the exclusion, coercion, and interrogation of victims.

Chapter 6 concludes the SAEK's story by turning to a broader discussion of technoscience, law, and sexual violence. Reflecting back on the SAEK's troubled history, the chapter considers what went wrong and why despite the decades of legal reform, rapid changes in forensic technology, and developments in expertise and training in sexual assault forensics and treatment, many of the same issues that plagued rape cases in the early 1970s persist. I describe advocates' visions of better futures for victims and for medicolegal practice, and consider the ways that the SAEK and its network may be reassembled in years to come.

2 Inscriptions of Doubt: Law, Anti-Rape Activism, and the Early SAEK

Rape is too serious a crime to simply accept a woman's word.
"OPP rape report prepared impartially" (1979)

No longer will we cry softly, padded by the courtroom walls.
We will scream out our collective fear, anger, and rage.
K. Zook (1980, 27)

The first Ontario Sexual Assault Evidence Kit emerged in the late 1970s, after a decade of mounting feminist anti-rape activism, vibrant public controversies about medical and legal responses to rape, and ongoing legal practices of distrusting and dismissing women reporting rape. These intersecting historical currents laid the ground for the SAEK's development. They inspired rape crisis advocates and medical and legal professionals to imagine a new technology that could serve as a techno-scientific witness of rape – a technology that many thought would offer more credible legal testimony of rape than that of the women who reported it.[1] In this chapter, I sketch these histories to chart how the kit came to be. I locate the kit's design in particular actors, practices, controversies, and beliefs about rape, victims, and credibility to reveal the politics of the SAEK's design. This chapter takes up the questions *who designed the Sexual Assault Evidence Kit, why, and how?* These deceptively simple questions open the door to a multilayered story of actors building expertise, alliances, and new technology and practices, amidst controversy and political strife.

By beginning with the question of *who* designed the SAEK, I start this trek into the kit's early history by putting the category of "the designer" at stake. In his well-cited text *Science in Action*, Bruno Latour (1987) argues that it is possible to see how science and technology are made by following scientists and engineers in action. Adhering to this line of thinking could have led me to follow the historical traces of forensic scientists and technicians who, in the late 1970s, worked to assemble a technological object for evidence collection. However, the kit's history demands a far more complex picture of technological design, one that takes into account the many different actors both within and outside of medical and legal institutions who shaped the contexts in which the kit came to be. Latour's directive to follow scientists and engineers has been widely criticized by feminist scholars who argue that this approach produces analyses that narrowly focus on powerful actors and ignore marginalized or less visible actors involved in scientific practice and technological design (Clarke & Montini, 1993; Star, 1991; Star & Griesemer, 1989; Wajcman, 2000). Being mindful of this potential, here I do not focus merely on the scientists and technicians who were at the SAEK's design table, or the physicians and lawyers who were responsible for collecting and interpreting the type of forensic evidence that the SAEK would later contain. Doing so would be to examine a predominately male cast of relatively powerful actors who were deeply entrenched in medical and legal institutions.[2] Instead, I expand my sightline to include feminist activists outside of medical and legal institutions, whose political struggles shone a light on sexual violence and the institutionalized forms of discrimination against sexual assault victims in Canadian hospitals, police stations, and courtrooms. By putting these actors at the centre of my story about the SAEK's design, I argue that their political work in and outside of medical and legal spaces was central to the SAEK's design. It played a pivotal role in building a context in which a new technology for evidence collection seemed necessary and inevitable.

Before the women's movement of the 1960s and the feminist anti-rape movement in the 1970s, rape and sexual assault had not entered into public consciousness (MacKinnon, 2005; Rutherford, 2011). Rape and sexual assault were commonly assumed to be rare crimes directed at women who placed themselves at risk. In the 1940s and 1950s, rape was rarely featured in academic literature and there were no protocols or formalized training in law enforcement organizations or hosptials dedicated to rape response (Dubinsky, 1993; Gavey, 2005). Before the SAEK could be imagined in the late 1970s, a shift in public consciousness about rape had to occur. In order to inspire changes in practice and developments of new technologies, rape had to first be politicized and become a crime that mattered politically. Importantly, expertise on rape and victim advocacy had to be developed by a group of actors before they could claim to know how the SAEK should be designed and why it was needed. The anti-rape movement in the early 1970s was the spark that ignited many of these changes. As anti-rape activists developed new expertise and increased public awareness on rape and rape response, they generated controversy and public dialogue about the failures of medical and legal institutions to respond to rape victims, which would later inspire the design of the SAEK.

Revealing the feminist anti-rape activists' work that preceded the development of the SAEK allows for a more multifaceted story about its design to emerge. The kit becomes a technology that materialized out of political struggle, activism, and controversies over practice and expertise. In the years preceding the kit's design, the question of *who* could claim to know rape and the best practices for rape response was a source of tension between anti-rape activists, police investigators, and hospital staff, which had to be negotiated. This context informed who could sit at the kit's design table and what role they could serve in constructing this new technology. This chapter paints a complex picture of the SAEK's design by charting the controversies in rape crisis centres, hospitals, and government consultations, which prompted the design of a new technology that many hoped would address the problems in sexual assault law and medicine that anti-rape activists were bringing into view.

Technological design involves more than just assembling material objects. Instead, as Akrich (1992) argues, it involves "inscribing ... vision[s] of the world in the technological content of the new object" (208). Parnis and Du Mont (2006) have argued that values of technical rationality and rape mythology are embedded in the SAEK's design. Here, I expand on this argument by illustrating the contexts that set the stage for the SAEK's design and the legal histories, practices, and politics that would later become inscribed in the kit's material form. I show how the kit emerged out of anti-rape activists' ongoing efforts to problematize and transform medical and legal practice in the 1970s, and illustrate the faith that many had in technology to address the biases in law and medicine, and produce evidence of rape that would appear more objective in court. I draw on Donna Haraway's (1997) understanding of the "modest [scientific] witness" (32) to discuss how the objectivity of medical forensic practice was upheld and challenged, and the promise that the SAEK seemingly held to make physicians and the evidence they produced more objective in the eyes of the law. Through this narrative, I reveal the historical politics of the kit, and its ties to the systemic doubt of victims' reports of rape in medicine and law.

Locating the SAEK's design in particular places, people, and contexts sketches the kit as a deeply contextual technology that bears the traces of its designers. Suchman (2003) contends that technological design is always "located" (4) and that detaching technologies from their sites of production runs the risk of making them appear inevitable. The kit's design was not inevitable, but was instead a product of negotiation and struggle. Locating technological design in actors, practices, and controversies, according to Suchman, demands a renewed responsibility for what the objects are and how they are used in practice. In order to understand how the kit works in contemporary practice, and demand a renewed responsibility for its many harmful effects on victims, it is crucial to understand the relations out of which the SAEK was first imagined, designed, and assembled.

Institutions and Rape

Institutional responses to rape in the 1970s drew together practices in law, medicine, and science, and, accordingly, the assumptions about women's credibility, sexuality, and experiences of violence that were embedded within them. These three institutions have historically carried similar cultural standing as neutral, objective systems of knowledge that can produce reliable truths (Epstein, 1996; Jasanoff, 1995). However, much of the feminist writing on law, medicine, and science have highlighted how these institutions have historically ignored, dismissed, and disqualified the experiences of people in marginalized groups, and in so doing, have produced truths that reflect patriarchal, racist, homophobic, and classist ways of knowing (Harding, 1991, 2008; Keller, 1985; MacKinnon, 2005; Smart, 1989). While medicine, science, and law similarly pose as objective and neutral, these scholars have argued, they *im*pose ways of knowing that reflect and reproduce white, male privilege and power. Aiming this critique squarely at the law, Smart (1989) argues that law is "a signifier of masculine power" (2) that resists and excludes women's everyday lives. In law and legal practice, she contends, women's everyday experiences are translated into legal frames that redefine and disqualify many aspects of their everyday lives. In doing so, the law asserts its power by claiming truths that deny the possibility of other truths and knowledge. Smart points to rape trials as an exemplar of how the law translates women's experiences into legal discourse, and argues that much of women's experiences of sexual violence are lost in the process. Wilkerson (1998) draws a similar conclusion about medicine, which she suggests forms an alliance with scientific ways of knowing to transform violence against women into an individual pathology that has no social origin. Taken together, Wilkerson and Smart point to how law and medicine translate women's experiences of rape into frames that exclude much of their lived realities of violence.

The design of the SAEK involved constructing a technology that translated victims' experiences of rape into forensic terms. But long before the SAEK, victims' experiences of rape were translated in the criminal justice system into medical, scientific, and legal language that often discounted and dismissed many aspects of their experience. In the early twentieth century in Canada, lawyers, judges, and police investigators worked alongside medical practitioners and forensic scientists to scrutinize victims' bodies for signs of sexual interference and to assert, contest, and claim legal truths about rape (Backhouse, 2008; Dubinsky, 1993). In this process of translation, victims' experiences of rape were put into language that was shaped by dominant, paradoxical views of women's sexuality and patriarchal and racist legal histories of viewing women as forms of property (Clark & Lewis, 1977; Gavey, 2005).

Women were viewed as forms of private property owned by their fathers and husbands in early British law (Clark & Lewis, 1977). Clark and Lewis argue that in this legal system, women were valuable as forms of property to the extent that they maintained their reproductive capabilities and sexual purity. Raped women, in the eyes of the law, were stripped of their sexual innocence and, therefore, their value to their fathers and husbands. Rape was thus a crime of theft of male property. This legal history informed early medical and legal practices that placed the victim's body and the damage that the rapist caused her body at the centre of rape investigations and trials.

Dominant views of women's sexuality also shaped early rape investigations and court proceedings. Gavey (2005) argues that women, particularly white women, have historically been paradoxically viewed as sexually passive beings with the dangerous potential to provoke male sexuality. This contradictory construction of female sexuality, she suggests, helped to push the assumed responsibility of rape from male perpetrators onto female victims, and propelled myths about rape being a female fantasy. While white women were seen as being simultaneously sexually passive and provoking, Davis (1983) argues that women of colour were viewed as inherently sexually promiscuous. This view of women of colour similarly displaced the responsibility of rape onto victims, and also fostered the widespread denial of rape against women of colour (Davis, 1983). These understandings of female sexuality fuelled centuries of medical and legal institutions dismissing and denying women's experiences of rape as being either a female fantasy or a victim's fault. Throughout this history, only those rapes that were physically violent and committed by a stranger against a woman who was believed to be sexually chaste were recognized as "real rape" in law (Estrich, 1986; Gavey, 2005).

Victims reporting rape to police in Ontario in the 1970s thus entered a system with a long history of dismissing, minimizing, and disregarding women's experiences of rape. They faced medical and legal institutions that had few, if any, services dedicated to rape response. With little training, few protocols, and no specialized services, police investigators and hospital staff had little to offer victims. The attention many victims received was inadequate and re-traumatizing (Clark & Lewis, 1977; LeBourdais, 1976; Williams & Williams, 1973). Common beliefs about rape as being a fantasy or a woman's fault filtered into practices in courtrooms, police stations, and medical exam rooms. Victims were often met with suspicion, judgment, and blame when they reported their rapes to police and hospital staff (Clark & Lewis, 1977). These practices in law, science, and medicine inspired the anti-rape activism that sought to resist them and laid the ground for the development of the SAEK as a credible technoscientific witness of rape.

Credibility, Distrust, and the Corroborative Evidence Doctrine

Reflecting early legal definitions of rape as a physically violent crime committed by male strangers against women, in the 1970s, rape in the Criminal Code of Canada (hereafter the Criminal Code) was narrowly defined as sexual intercourse perpetrated by a male against a female who was not his wife, without her consent or with her consent obtained through force, fear, or fraud (s. 143). Under this narrow definition, men and boys could not be rape victims. And if a woman was raped by her husband or by means other than forced vaginal penetration, her experience was not legally recognized as rape. Laced into the inherently gendered and sexed definition of rape in law was a history of sexist legal practice premised on beliefs about women's inherent lack of credibility.

In the 1970s, rape victims were commonly seen in law and legal practice as untrustworthy and lacking in credibility (Backhouse, 2008; Dubinsky, 1993; Martin, 2005; Taslitz, 1999). If rape is a female fantasy, so the dominant reasoning went, then women could not be trusted to tell the truth about rape. Women reporting rape were often doubted by police and their credibility as witnesses in rape trials was under constant challenge (Clark & Lewis, 1977). A woman's race, class, age, and marital status all influenced the extent to which her report of rape was likely to be believed (Pierson, 1993). In rape cases, as in other criminal cases, the victim served as a witness for the Crown. A female rape victim's credibility as a witness was often put on trial in the courtroom. Defence lawyers commonly challenged a woman's credibility by digging into her sexual history and arguing that a woman's sexual past implied her consent to all sexual acts, indicated her diminished moral character, and suggested that her testimony was likely false (Clark & Lewis, 1977).[3] This practice had deep roots in nineteenth-century English law, where a woman's unchaste character was often used in court to demonstrate her consent to a sexual act and her lack of credibility as a witness (MacFarlane, 1993) – a view that was later solidified in Wigmore's (1940) influential treatise on evidence, which stated that "no judge should ever let a sex-offence charge go to the jury unless the female complainant's *social* history and mental make-up have been examined" (737, emphasis added). A woman's social history determined the extent to which she was considered a reliable and credible witness. A Crown attorney who worked in Ontario courts in the 1970s described how this sentiment fed into legal practice at the time. For a woman's testimony to be believed, he said, "you had to be practically a virgin who was kidnapped from the convent."[4]

The significant mistrust of women reporting rape shaped police investigations as well. It was common practice in the 1970s for police to dismiss women's reports of rape as unfounded, a designation that was reserved for cases where it was believed that no violation of the law took place or was attempted (CBC, 1971; Police crime statistics, 1975). In 1975, Ontario police dismissed 40% of reported rapes as unfounded, as compared to 6.5% of assault reports, 8% of robbery reports, and 5.9% of breaking and entering reports, and laid charges in only 30% of rape cases (Police crime statistics, 1975). As in the courtroom, a woman's social location, martial status, and sexual history influenced the likelihood she was to be believed by investigating officers. Clark and Lewis's (1977) detailed analysis of rape investigations revealed that police were significantly less likely to believe a woman's rape report and lay charges if she had been drinking or hitchhiking, was under nineteen years of age or over thirty, had no injuries, or was separated or divorced (Clark & Lewis, 1977).

The prevailing view of women as unreliable witnesses fuelled legal demands for independent corroborative evidence of rape (Backhouse, 2008; Parnis & Du Mont, 1999). Sir Matthew Hale, a seventeenth-century judge, is often credited with penning one of the first legal arguments for the necessity of corroborating reports of rape. He wrote: "It is true, rape is a most detestable crime ... but it must be remembered that *it is an accusation easily to be made,* and hard to be proved, and harder to be defended by the party accused, tho never so innocent" (as cited in MacFarlane, 1993, 53; emphasis added). Hale's words became a dictum that provided the legal justification to doubt women's credibility in rape cases and invited the legal requirement for corroborative, independent evidence that could prove the veracity of her testimony (MacFarlane, 1993). Three centuries later, Hale's doctrine continued to be a powerful influence in English and North American common law, as was evident in the following words of one legal scholar in the late 1960s: "Surely the simplest, and perhaps the most important reason not to permit conviction for rape on uncorroborated word of the prosecutrix is that the word is very often *false* ... Since stories of rape are frequently lies or fantasies, it is reasonable to provide that such a story, in itself, should *not* be enough to convict a man of a crime" (Corroborating charges of rape, 1967, 1138; emphasis added). Echoing Hale's argument, the scholar argued that rape cases required more corroborative evidence than other offences because rape was "uniquely difficult to disprove" (1139).

Beliefs about the importance of independent corroborative evidence were encoded in Canadian criminal law. Canadian judges in the 1970s followed a corroboration rule for rape cases that was explicitly laid out in the Criminal Code, which stated that in the absence of corroborative evidence that implicated the accused, judges should warn juries about the dangers of convicting an accused on the basis of a victim's testimony alone. Although this instruction was eliminated from the Criminal Code in 1976, after significant lobbying from anti-rape activist groups, it did not significantly change legal practice (Clark & Lewis, 1977; Du Mont & Parnis, 2001; Osborne, 1984). Using common law as justification, judges continued to warn juries about the dangers of convicting the accused in rape cases without corroborative evidence, until 1983, when more significant legal reforms prohibited these warnings (Osborne, 1984). Requiring corroborative evidence in rape cases, according to Backhouse (2008), was "a substantial deviation from the general principles of evidence" (171), which in most other cases, allowed jurors to determine a witness's credibility on the sole basis of their testimony in court and not on the basis of independent evidence or the testimony of another. The additional evidentiary requirement of independent corroborative evidence in rape cases in the 1970s made them far more difficult to successfully prosecute (Backhouse, 2008).

Injuries and signs of force on a victim's body were often deemed to be corroborative evidence of rape (Parnis & Du Mont, 1999; Du Mont & Parnis, 2001). While there was nothing in the Criminal Code that required evidence of force for a rape conviction, Backhouse's (2008) historical research suggests that "judges and juries were historically loath to convict without evidence of substantial force and spirited resistance" (147).[5] This practice had a long history in Canada. At the turn of the century, Canadian medical doctors routinely conducted forensic medical examinations on rape victims and gave testimony in court on whether corroborating physical evidence of force and traces of semen and sexual diseases were present (Dubinsky, 1993). Medical textbooks at the time instructed physicians to apply scepticism to these medical exams of rape victims because, as some texts asserted, rape was easily avoidable with physical struggle and false allegations of rape were "frequently made for the gratification of malice and revenge" (as cited in Backhouse, 2008, 40). These texts advised physicians to not only observe the victim's vaginal area and their undergarments for signs and traces of penetration, but also their walk, attitude, mental capacities, and other bodily injuries to access the validity of their reports of rape (Backhouse, 2008). The sexism, classism, and ableism embedded in these texts shone through the instructions to physicians to be more suspicious of working-class victims and victims who did not physically resist the attack to the extent that physicians deemed appropriate and natural. According to Mills (1982), nineteenth- and twentieth-century medical examiners saw their role in rape trials as supplying objective truths against women who routinely lied about rape.

In the early 1970s, some physicians were beginning to acknowledge in medical journals that rape was not always physically violent (e.g.,

Burgess & Holmstrom, 1973); however, the evidentiary requirement of force and injuries held strong. Without medical evidence of injuries, convictions were extremely unlikely (The rape corroboration requirement, 1972). Like the doctors at the turn of the century, physicians in the 1970s conducted medical forensic exams on victims to document signs of violence and traces of semen that could corroborate a woman's testimony. Forensic scientists in crime laboratories used an array of technologies to analyse the evidence medical doctors collected. Through this work, practices in medicine, science, and law transformed the victim's body into a site for testing the validity of reports of rape. What physicians saw on the victim's body was taken in law to reveal truths of rape; whereas what victims said about their rape was taken in law to be highly suspect. Armed with the apparent objectivity and neutrality of medical science, physicians' testimony had the credibility in court that victims did not. The institutions of science, medicine, and the law thus coalesced in the Canadian criminal justice system to eternalize and re-enact legal histories of distrusting women reporting rape.

Science, Medicine, and Technoscientific Proof of Rape

Prevailing views of forensic medical evidence as inherently *objective* in law fit neatly alongside the distrust of women's testimonies of rape. Scientific objectivity, many feminist scholars have argued, has always been gendered (Haraway, 1997; Harding, 1991; Keller, 1985). As I discussed in chapter 1, Haraway (1997) describes how masculinity came to be associated with objectivity in the beginnings of experimental science in the seventeenth century. The male scientist was seen as a "modest witness" (32) of the natural world who gained his credibility and objectivity from his presumed capacity to make himself invisible in his scientific reports by transcending his body and operating purely in the mind. Women and working-class men were assumed to be unable to achieve this form of modesty, and were thus unable to make objective, credible, scientific statements about the natural world. In the Canadian criminal justice system in the 1970s, the physicians and forensic scientists who collected and analysed medical forensic evidence in rape cases were presumed to be the *modest witnesses* of rape, who, unlike the female victim, could transcend their bodies and personal biases to produce objective facts about rape. While the rape victim was confined to speaking from her body, these actors seemingly had the power to operate purely in the mind. This faith in forensic science and medicine to reveal objective truths about rape played out in forensic laboratories, courtrooms, and police stations.

The forensic evidence that physicians and forensic scientists collected and analysed was itself imbued with authority. In the 1978, Dr Krishnan, a scientist at the Ontario Centre of Forensic Sciences (CFS), argued that forensic science in the 1970s had been "propelled into the modern age" (5) with new technologies that offered higher levels of magnification and precision in chemical analysis of bodily fluid samples.[6] He praised the forensic evidence being produced in laboratories of the 1970s for being impartial, "objective by nature" (9), and "not subject to lapse of memory, confusion, and perjury, as are human witnesses" (11). These beliefs in the objective powers of forensic evidence crept into the court-room. The semen, hair deposits, and drugs in the victim's blood stream, along with the medical evidence of scratches and bruising on victim's bodies, were taken as necessary and objective indicators that a rape had indeed occurred (Training manual, 1977). This kind of medical foren-sic evidence was, according to historical records from rape crisis cen-tres, "almost a prerequisite for conviction" (Present laws, ca. 1980, 5). A retired defence lawyer who worked in Ontario courts in the 1970s recalled in an interview how "science was worshipped" in the court-room and forensic scientific evidence, when it was available, was taken "as gospel." In the eyes of the law, forensic evidence was, in the words of a Crown attorney at a rape awareness conference in 1975, "necessary to prove rape or indecent assault on a woman" (One day conference, ca. 1975). Forensic medical evidence, unlike human witnesses, was seem-ingly untouched by bias and had the capacity to speak objective truths about rape. Along with the physicians and scientists who collected and analysed it, this evidence was treated in court as a credible, objective *witness* to the facts of the crime.

Under the force of these dominant beliefs about the powers of foren-sic evidence, victims of rape were encouraged to go to hospital emer-gency wards for forensic medical examinations. For many rape victims, treatment in these wards in the 1970s was often insufficient and trau-matizing (Burgess & Holmstrom, 1973; Clark & Lewis, 1977; Williams & Williams, 1973). In one study, one-third of the victims who visited Ottawa hospitals in 1976 reported feeling frightened and intimidated by the medical professionals who treated them (LeBourdais, 1976). Some women reported feeling just as traumatized by their medical treat-ment as they were by their rape (Donadio & White, 1974). The general disinterest in addressing women's health issues that was so prevalent in medical institutions at the time (Murphy, 2012), along with popular notions of rape as being a woman's fault, shaped hospital staff's responses to sexual assault victims. Reports documented victims' common experiences of physicians discounting their assaults and making comments such as "There is a fine line between rape and promiscuity" (Kinnon, 1981, 27), and "There is no such thing as rape" (LeBourdais, 1976, 12). One victim reported that her gynecologist admonished her by saying that "a hospital should not be considered to be a counseling service or a haven for the unloved or the unwashed" (ibid., 102).

Institutionalized forms of racism and classism shaped the care that victims received. Archival records indicate that a victim's race and class had great bearing on how physicians and nurses treated her and the likelihood that they believed her story (Training manual, 1977). Several victim advocates that I interviewed recalled that many physicians and nurses treated middle class, married, white women as "one of their own," whereas racialized, impoverished, sexually active, young women were more commonly treated as either deserving or lying victims. Reflecting this trend, a Toronto gynecologist was quoted at the time as saying, "There is one type of woman I would have a hard time believing was raped: a woman between 16 and 26, on the pill and no longer a virgin" (LeBourdais, 1976, 102).

Despite the value that was placed on medical forensic evidence, there were few standardized protocols for medical evidence collection in rape cases (Hargot, 1982; Williams & Williams; 1973). Some standardized protocols for sexual-assault forensic evidence collection were beginning to appear in the United States in the early 1970s (Evrard, 1971; Fahrney, 1974); however, there are no records of similar protocols in Canada in the same period. Evidence collection in Ontario rape cases was, according to one physician, "dealt with in a haphazard fashion" (Hargot, 1982, 126). A forensic scientist I interviewed said that without consistent procedures within or across Ontario hospitals, physicians had unregulated discretion to collect medical evidence with whatever materials they saw fit. Spare swabs, vials, and envelopes in the emergency ward were likely used for evidence collection when individual physicians deemed these procedures appropriate. The lack of protocols for evidence collection was accompanied by lack of forensic training for physicians and nurses (Kinnon, 1981). Without much understanding of forensic evidence collection and preservation, risks were high of samples moulding and being contaminated before they arrived at the forensic lab.

Insufficient training and a lack of forensic protocols weighed heavily on many victims in emergency wards (Donadio & White, 1974; Kinnon, 1981). Lengthy delays and extended examinations were common. Many victims also reportedly found forensic exams to be very intrusive, painful, and traumatic (Kinnon, 1981). Exams usually included physicians plucking a victim's pubic and head hairs, clipping her fingernails, drawing her blood, photographing her injuries, and conducting pelvic exams and vaginal aspirates (One day conference, ca. 1975). One victim described how invasive these procedures were: "What he was doing was so similar to what just happened ... it was too much the same, two times in one night ... I didn't want anyone to look at me; I didn't want anyone to touch me" (as cited in Kinnon, 1981, 26).

Many physicians in Ontario were reportedly reluctant to conduct medical forensic exams on rape victims (LeBourdais, 1976). Although these medical forensic exams seemingly brought the disparate worlds of science, law, and medicine together, physicians' reluctance to conduct them marked one of the many tensions brewing within this coalition. Ontario physician Dr Len Hargot (1982) explained that physicians' resistance to the forensic exam stemmed from their discomfort with becoming involved with legal proceedings, their lack of training in forensic procedures, the minimal financial compensation they received for the exam and time in court, and the time the exam required. As a result, many victims were turned away from emergency wards and, in the words of one report, "forced to shop around for medical treatment" (LeBourdais, 1976, 103). One hospital administrator explained this practice of turning victims away by saying that "the hospital "doesn't encourage" rape victims to come in for examinations because doctors lose too much time when cases get to court" (as cited in Kinnon, 1981, 26). These delays and refusals to conduct forensic medical exams, along with the lack of protocols and forensic training, made many victims' path to securing the seemingly necessary corroborative forensic evidence extremely difficult. Many were thus caught between institutions of law enforcement where their reports of rape were not believed without scientific proof and medical institutions that provided unreliable services for evidence collection.

Resistance to these harmful and ineffective medical and legal responses to rape was stirring in the early 1970s. Growing out of the women's liberation movement of the 1960s, a feminist anti-rape movement in Canada was on the rise. The activists in this movement sought to define new expertise and knowledge about women's experiences of rape that had historically been excluded in law, and draw attention to the sexism of law, science, and medicine that played out in rape investigations and trials. Their work would significantly shape the landscape of medical and legal responses to rape and play an important role in initiating the development of the Sexual Assault Evidence Kit.

Anti-rape Expertise and the Rise of Rape Crisis Centres

In the late 1960s and early 1970s, the North American women's movement was in full swing. Drawing its momentum from social justice movements in the 1960s, the women's movement was developing language and analysis of patriarchy and gender-based oppression (Bumiller, 2008; Cohen, 1993; Corrigan, 2013a; Pierson, 1993). By 1966 in Canada, the Equality of Women in Canada Committee had formed a Royal Commission on the Status of Women, which developed 167 recommendations that were aimed at "ensur[ing] for women equal opportunities with men in all aspects of Canadian society" (Report of the Royal Commission on the Status of Women, 1970, 13). New feminist literature had inspired feminist consciousness and the birth of feminist organizations across North America (e.g., Friedan, 1963; Millet, 1969). Many women's liberation groups emerged in Canada in the late 1960s, such as the Toronto Women's Liberation Movement, the Fédération des femmes du Québec, and the Vancouver Feminine Action League. Women and their allies were organizing women's caucuses in labour unions, instigating collective childcare, establishing safe houses for women fleeing from violence, and organizing to legalize abortion (Rebick, 2005).

Out of this diverse movement, Canadian women formed consciousness-raising groups and shared their experiences of gendered oppression and violence at the hands of men (Cohen, 1993; Kinnon, 1981; Vance, 1978). Speak-outs provided a public forum for women to share stories of violence and a public face of the anti-rape movement in the 1970s (Kinnon, 1981; Vance, ca. 1977).[7] This work, according to activist Susan Cole (1989), "uncovered the truth that sexual abuse was epidemic, not occasional, [and] more normal than marginal" (12). With the release of Susan Brownmiller's *Against Our Will: Men, Women, and Rape* in 1975 and the first Canadian book on rape in 1977, Lorenne Clark and Debra Lewis's *Rape: The Price of Coercive Sexuality*, the anti-rape movement in Canada was giving voice to an issue that had long been silenced within and outside of medical and legal institutions.

Anti-rape activists rallied around radical feminist analyses of rape as an expression of patriarchy, sexism, and women's oppression (Cohen, 1993; Neigh, 2012). American feminist Susan Brownmiller's (1975) assertion that rape was an act "by which all men keep all women in a state of fear" (4) epitomized this analysis and was commonly cited in writings by Canadian anti-rape activists (e.g., Toronto Rape Crisis Centre, 1979; Winner, 1977). Rape, it was argued, uniformly cut across lines of class, race, ability, and sexuality and threatened all women's safety: "Women from all walks of life are raped. No woman is safe from the possibility of sexual assault" (Toronto Rape Crisis Centre, 1979, 3). This argument confronted and repudiated views of rape that had been circulating in medicine and law for centuries. Rape was not a female fantasy, a rare occurrence, or the fault of victims, these activists argued. Rather, it was a systemic social problem that affected all women. This analysis glossed over the different experiences of rape among women of colour, Indigenous women, and disabled women, as many feminists would later argue.[8] However, by drawing attention to the systemic nature of rape, these activists were defining new understandings and truths about rape that had been denied and rebuffed in medical and legal institutions.

In addition to building new knowledge about rape itself, anti-rape activists were assembling knowledge about victims' experiences in the justice system. They began to see how common it was for victims to feel revictimized by hospital staff, police, lawyers, and judges and how serious the need was for supportive services for victims. Reflecting the anger that accompanied this new consciousness, anti-rape activist Dorthi Winner (1977) wrote: "For too long, a woman's victimization through the crime of rape has extended to her being victimized through the predominantly male and sexist-oriented world of medical examinations, interview with police, and the blatantly injust [sic] court system. For too long, women have been raped over and over again by these processes" (1). Medical and legal institutions were not sites of objective, neutral practice, many activists argued. Instead, practices in these institutions were laden with sexism. "Canadian rape laws reflect patriarchal attitudes and discriminate against women" (Legal policy of the National Association, 1978), one group of activists wrote. Echoing this critique a few years later, anti-rape activist Joni Miller (1981) described the discrimination and misogyny that marred forensic exams and police investigations of rape:

The treatment a woman receives from the police can vary widely according to who she is, what she looks like, and the particular attitudes of the
particular police officer. She could be treated with consideration or
she could be asked if she enjoyed it and what she did to entice the man ...
The police pathologist arrives at the hospital ... [and] after being forced
into sexual intercourse with one man, she is required to submit to another
man to validate what she says happened to her. (5)

Police investigators and hospital staff, in the eyes of many anti-rape
activists, were unable to transcend their personal biases, and were thus
far from the objective *modest witnesses* of rape that they were assumed
to be. Instead, activists argued, these actors' practices were profoundly
shaped by the sexism and misogyny of medicine and law.

In response to this climate, anti-rape activists began to organize rape
crisis centres (RCCs) in the early 1970s that aimed to provide supports
and advocacy for rape victims. RCCs were intended to be "feminist
counter institutions" (Murphy, 2012, 191), along with the feminist
health clinics, women's bookstores, and battered women's shelters that
were opening around the same time. Many RCCs became hubs of feminist political action and anti-rape public education, feminist consciousness raising, and expertise on feminist peer counselling and anti-rape
advocacy. Activist Susan Cole (1989) later recalled that in addition to
the supports RCCs offered victims, "these centres took on what seemed
like the overwhelming task of educating the public and smashing the
ancient and very resilient myths about violence against women" (12).

The first RCC in the United States opened in 1970 and not long after,
Vancouver Rape Relief opened its doors in 1973 and the Toronto Rape
Crisis Centre (TRCC) in 1974. One of the first advertisements for the
TRCC described the centre as "a place where women who have been
raped can go and be helped ... where she can sit and talk with sympathetic women, [and] have a warm drink" (Rape crisis centre opening,
ca. 1974, 1). The pressing need for this service was evidenced in the
first year, during which the TRCC answered 2600 calls from Toronto
women seeking advice and support, a number which increased rapidly in the following years (Brief submitted, 1974). Other rape crisis
centres opened in Ontario and across Canada, some with activists running 24-hour crisis lines out of their homes and others renting space
in community buildings. With primarily volunteer labour, these centres offered counselling, advocacy, and victim accompaniment in hospitals, police stations, and courtrooms. Through this work, anti-rape

activists developed a body of feminist expertise on victim advocacy that emerged from women's experiences of rape and the criminal justice system. Writing about these beginnings, the Ontario Coalition of Rape Crisis Centres (1991) recollected: "There were no guidelines to draw from, other than our own and other women's experiences. We were faced with the necessity of developing our own training materials as well as developing expertise required to provide help to many different women" (1).

Many of the Canadian RCCs in the 1970s saw themselves as taking on a "dual role, as a woman-oriented alternate social service, and as a vehicle for social change" (Vance, ca. 1977, 1). While some scholars have argued that RCCs were often forced to choose between service provision and political action (Cohen, 1993), early descriptions of RCCs' work suggest that some RCCs in the 1970s were collapsing distinctions between service provision and political work (Training manual, 1977). Some centres offered feminist peer counselling, which resisted medicalized definitions of therapy that they saw as pathologizing women's experiences of rape, and insisted instead that women's experiential knowledge could be the basis of support for other women (Marriner, 2012). One advocate from the Toronto RCC explained that "the principle of peer counseling with a feminist approach was revolutionary ... Every feminist service was a reaction to patriarchy" (Parent as cited in Rebick, 2005, 82). Another described RCCs' work as being not about service but about "solidarity and self aid" (Lakeman as cited in Rebick, 2005, 73). This work, for many activists, was propelled by optimism that rape crisis advocacy could combat and eventually eliminate male violence. An activist I interviewed who worked at an RCC at the time recalled, "We all believed that we could put ourselves out of business."

Many RCCs adopted models of organizing from the women's liberation movement and other social justice movements. They operated as collectives in opposition to the hierarchical organizational structures, which they saw characterizing medical, legal, and many other social institutions (Cohen, 1993). One activist described the collective model her centre used by saying, "The whole idea was to share the work, rotate the authority, and follow emerging leadership" (Lakeman in Rebick, 2005, 71). Some centres, largely as a result of pressures from funders, moved quickly to instituting a hierarchical model with a board of directors, but most in Canada maintained their collective model for much of the 1970s. By 1977, there were eleven RCCs across Ontario, and an Ontario Coalition of Rape Crisis Centres (OCRCC) had been initiated. In 1978, a National Network of Rape Crisis Centres was formed (Minutes of AGM, 1978). With this collective strength, many anti-rape activists in RCCs turned their attention to challenging and transforming medical and legal practice.

Relations with Institutional Actors

Rape crisis centres in the 1970s had a complicated relationship with medical and legal institutions. While archival government records commonly depict RCCs as non-political service providers that worked collaboratively alongside doctors, police, and legislators (see Provincial Secretariat for Justice, 1979a), these portrayals conceal the tensions, struggles, and power differentials between many anti-rape activists and the institutions they sought to change. While some activists in the movement were debating RCCs' relationships with hospitals and law enforcement organizations, others were fighting to gain entrance into institutional spaces to reform law and medical and legal practices in rape cases (Fitzgerald, 1982; Zook, 1980). Gaining entrance into these institutions was often a strategic affair, according to many of the activists I interviewed. Activists described how they struggled through institutional barriers to fight for rape law reform, train police and physicians to respond more sensitively to victims, and advocate for victims in hospitals and police stations. To be taken seriously by doctors, police, and legislators, some made calculated choices to downplay the feminist politics of the movement and assert themselves as legitimate experts – actors who belonged in medical and legal networks of rape response. Describing this struggle, one activist said that they had to "fight to prove that [they] were needed and that [they] were credible." To do so, they translated their expertise, born from their experiences of supporting victims of rape, into terms that institutional actors would be willing to accept. Tensions often ensued when physicians, police, and legislators fought to protect the boundaries of their work with policies and practices that pushed RCC advocates to the margins of institutional practice. These struggles and controversies shaped the context within which institutional actors and RCC advocates later imagined and designed the SAEK and set the stage for the complicated dynamics between the kit's designers. Given their significance to the kit's design story, the struggles between anti-rape activists and institutional actors that played out in law reform consultations rooms, training sessions, forensic exam rooms, and government boardrooms warrant some attention.

Through the 1970s and into the early 1980s, there were ongoing debates in the anti-rape movement about the efficacy of RCCs working collaboratively with hospitals and law enforcement organizations. While some activists saw collaboration with medical and legal institutions as a necessary route to enacting change, others felt it was in direct opposition to the core values of the movement. Controversies waged in feminist publications. In *Broadside*, a Canadian radical feminist newspaper, Zook (1980) wrote:

> Institutions exist only to reinforce those roles which perpetuate rape ... When we continue to try to change the institutions we slip into a pattern of upholding them in order to keep our access to them open ... Rape crisis centres are developing as traditional institutions by rationalizing that it is important to show doctors, lawyers, police and social workers how to better do their jobs ... This supporting of institutions institutionalizes rape as an accepted social reality. This is only adjusting, not facilitating value changes. (27)

For Zook, and many others, working alongside doctors, police, and legislators to reform institutional practice perpetuated the inequalities that bred violence against women. Others disagreed. Maureen Fitzgerald (1982) later countered Zook's view: "There is a growing and understandable impatience with the band-aid solutions ... [However,] when we put pressure on the state to provide services for women, we are putting the pressure where it belongs" (4). In these heated exchanges, the identity of the movement was constested, as activists debated not only their place in medical and legal networks, but also the terms of their activism, collective values, and strategies for enacting change. While these controversies continued, many RCC advocates ventured into institutional spaces, where the terms of their expertise and advocacy were in constant tension and negotiation with institutional actors.

One place where these negotiations occurred was during the consultations on rape law reform. In the late 1970s, anti-rape activists in Canada and across the United States were advocating for reforms to existing rape laws (Corrigan, 2013a; Sheehy, 1999). In 1975, anti-rape activists from across Canada met in Vancouver and Ottawa to draft recommendations for the Law Reform Commission, a committee charged with the responsibility of redrafting Canadian rape law (Vance, 1978).

The activists made twenty-four recommendations, the most significant of which proposed eliminating rape as a distinct offence in the Criminal Code and reclassifying it as a form of sexual assault (Preamble, 1975). Many activists hoped that this reform would increase the law's capacity to recognize the varying forms of sexual assault that women experienced and encode the law with feminist understandings of rape as an act that was not sexual, but instead inherently violent. It would be eight years until these recommendations would be formally adopted in law. During the preceding years, anti-rape activists across Canada actively lobbied the government to adopt their recommendations (Sheehy, 1999). They sought to lend credibility to their recommendations by situating themselves as experts on rape and demanded that legislators and law enforcement officials recognize them as such. One RCC report to the solicitor general stated,

> Rape crisis centres' staff and volunteers have acquired *considerable expertise* on the issue of rape and sexual assault. They have studied the law relating to the issue and have a *wealth of practical experience* in dealing with the implications of the law as well as implementation of the procedures and policies of institutions. Women who work in rape crisis centres are in a *key position* to recommend improvements in legislation concerning rape and sexual assault. (Relevance of rape crisis centres, ca. 1977, 103)

RCCs' involvement in the debate on rape law reform was not inevitable. *It could have been otherwise* – to use Everett Hughes's (1971) well-cited phrase – had it not been for the activists who worked to situate RCCs as sites of expert knowledge on rape. Halfon's (2010) understanding of expertise provides some useful insight here. He suggests that expertise is not a possession, but rather is situational and performative. To gain entry into law reform discussions, anti-rape activists had to perform their expertise in ways that appeared credible and non-threatening to legislators. This performance of RCCs' expertise played out in other sites as well, often with varying results that sometimes included direct conflict with institutional actors.

Some RCCs offered training for police, lawyers, and hospital staff on sensitive and appropriate treatment of rape victims, which was then the only professional training of its kind. One advocate I interviewed remembered how these educational sessions built on the trainers' "own experiences about what women needed." In these training sessions, activists asserted themselves as experts not on the basis of professional status or academic credentials, as many of the hospital staff, police, and lawyers did, but instead on their experiences of advocating for rape victims. Not all institutional actors welcomed these training sessions or the reconfigured definition of expertise upon which they were based. One retired Crown attorney recalled:

> There was a certain amount of strain between the women's groups and the law enforcement administration of justice ... I think there is always a tendency to say ... "You can't tell me that I'm not doing a good job." And to be fair to myself and the people that I worked with, there was a certain amount of maybe overzealousness, stringency in the special interest groups that were trying to generate change.

The accusation of stringency went both ways. Amidst the internal debates about the movement's relationship with institutions, some activists saw workplace training as an ineffective tool for change. Lee Lakeman, a member of Vancouver Rape Relief and the National Association of Rape Crisis Centres, explained, "At a certain point ... we stopped training cops and emergency room nurses, because these were black holes where you could be taken up forever" (as cited in Rebick, 2005, 75). In another interview she continued, "I think it is much more important to realize that these are hierarchies and what you have to do is affect the top ... You're not going to have much impact training the troops" (as cited in Neigh, 2012, 85). While some RCCs stopped offering training in institutions for these reasons, many others continued well into the 1980s.

The tensions between activists and institutional actors often played out in hospital emergency wards as well, when anti-rape activists would accompany victims as advocates in their forensic medical exams. In the exam room, RCC advocates had to negotiate their place in medical and legal practice and find ways to carefully present themselves to medical staff as non-threatening, yet knowledgeable about sensitive sexual assault care. A training manual for RCC volunteers from 1977 revealed how difficult a line this was for advocates to walk and described police and emergency room personnel as "people we can't afford to alienate" (Training manual, 1977, p. 28). It stressed the importance of challenging oppressive and harmful medical and legal practice, but also warned volunteers that antagonism towards medical staff could "backfire" (28) on victims. One RCC advocate recalled that finding this balance was often a challenge. She said, "We had a pretty combative relationship with the hospital and with the police. And while we offered advocacy to the hospital as an option, we did so with some trepidation, feeling that women would not be treated respectfully through that process."

RCCs trained their advocates to work as "buffers" (Training manual, 1977, 43) between victims and hospital emergency staff. Being a "buffer" involved not only responding to victims' needs, but also ensuring that medical staff offered appropriate and sympathetic medical care. RCC advocates were trained to assert their expertise in delicate ways, so as not to offend the physicians or nurses in the exam room. One RCC training manual cautioned advocates with the following instruction, "You may have to make suggestions to medical folk ... about what to do and how to do it. Be firm but gentle ... say it nicely, but do say it. You may be averting disaster" (43). Training manuals also instructed volunteers to remind physicians about their responsibility to be objective during the forensic exam: "The doctor will write a report – he is likely to be judgmental and add his opinions to the report – remind him that he need only report on the medical information" (Legal-medical procedures, 1983, 2). Volunteers were also instructed to advise physicians on how to use medical tools in less harmful ways for victims, such as the "request that the doctor warm the speculum for the examination" (2). Being a "buffer" between victims and hospital emergency staff required a careful performance of expertise.

Having training manuals that detailed the steps of forensic evidence collection, RCC advocates likely knew more about evidence collection than many of the physicians conducting it, who were receiving little to no formal training on forensic exams at the time. However, this knowledge rarely translated into authority. One advocate recalled how many advocates were hesitant to assert their specialized knowledge not only because of its potential to offend hospital staff, but also because of the differences in professional status, institutional power, and sometimes gender, race, and/or class between the RCC advocates and physicians. Within these complicated power dynamics, most often RCC advocates in the medical exam room were critical observers of medical practice. They became directly involved only during moments when they were able to follow the training manual's advice and gave physicians directions on evidence collection. If they followed the training manual, these advocates expressed their expertise in "gentle" and "nice" ways so as not to threaten or challenge physicians' medical expertise (Training manual, 1977). Physicians, however, had no obligation to follow advocates' instructions. The advocates were thus experts in the medical exam room whose expertise and directions could be dismissed and whose presence could be ignored. Through these tense relations, advocates were positioned both within and outside of medical practice. They were, to use Star's (1991) term, *marginal actors* who were insiders with specialized knowledge on evidence collection and, simultaneously, potentially threatening outsiders who had the potential to challenge and threaten medical expertise.

The tense relations that RCC advocates had with hospital staff were made more complex by the larger context in which RCCs were operating. Through the 1970s, RCCs were facing increasing financial pressures, which were forcing new relations with government institutions. As the non-renewable, short-term government grants that funded the opening of most RCCs began to expire in the late 1970s, many centres were struggling to cover the basic costs of rent, phone lines, and staff. At the National Day of Protest against violence in 1977, activists spoke about the funding shortages threatening RCCs: "We are faced with the loss of what little we have gained in the way of assistance to women … This is not accidental. It is systematic. Women, always and everywhere, have had to accept the burden of worsening economic conditions" (National day of protest, 1977, 1). In 1980, eleven out of the fourteen centres in the Ontario Coalition of Rape Crisis Centres were threatened with closure (Morton, 1980, Morton to unknown, 14 January 1980). With increasing media attention on pending RCC closures across the province, public pressure fell on the provincial government to provide RCCs with permanent funding.

The prospect of securing provincial funding raised difficult questions for many RCCs about the importance of their autonomy from formal institutions. Several of the anti-rape activists I interviewed remembered how RCCs at the time were debating the merits of becoming government-funded agencies. They recalled that although many advocates believed that accepting government funds was necessary to save RCCs from closing, the potential consequences of government agencies suppressing and controlling feminist politics and activism through funding agreements were widely discussed. The Canadian Rape Crisis Centres' (1979) funding manual warned RCC fundraisers, "If you accept the funds, you accept all the strings that are attached" (9). Considering this possibility, one Ontario activist wrote about her fears that government funding would force RCCs to become merely "appendages to the existing social service delivery system" (Vance, ca. 1977, 5). Government funding agencies, she argued, "have their own hidden agendas" (5) that would de-radicalize anti-rape activism in RCCs. Despite these hesitations, the increasing pressures from lack of funding led the Ontario Coalition of Rape Crisis Centres (OCRCC) to apply for $750,000[9] from the Ontario Provincial Secretariat for Justice (PSJ) in 1979.

The OCRCC funding application initiated a series of funding negotiations between the OCRCC and the PSJ. OCRCC representatives were faced with difficult decisions about the terms of their new relationship with the provincial government. One OCRCC fundraiser recalled that early on in these negotiations, it became clear that the government's primary interest was in gaining access to the information that centres had on women's experiences of rape. She said, "I think it was pretty clear to all of us that we were maybe going to have to do some things that we didn't want to do in order to keep the money … I remember we sat in each other's hotel rooms … into the middle of the night having those conversations … convincing each other that we were doing the right thing." The fundraisers strategized on how to make RCCs appear simultaneously credible and non-threatening to the government funders. The RCC fundraiser I interviewed described how she and the other advocates worked to present RCCs as non-political services for victims to the funders at the negotiating table. They cast themselves as service providers whose work the government had an obligation to support and tried to de-emphasize the feminist politics and activism within RCCs. She explained, "We rarely talked to them about it from a feminist point of view. And, if anything, we worked together. If we thought somebody was a little bit out of control, under the table, we would pat their hands. I remember doing that [laughs]." Beyond their selective choice of words, they also tried to embody the non-threatening, apolitical persona of RCCs that they were trying to portray. The advocates solicited advice from female politicians on how to dress; the RCC fundraiser I interviewed remembered being told, "If you turn them off by your appearance, they may not listen to what you have to say. If you are coming in to their world, you need to dress like you are in their world." She continued the story with laughter, "So we did. Yah, it's quite comical when you think back [laughs] … I remember borrowing gloves … from my mother. We wore gloves to the meeting! And you are talking about a group of women who didn't own a pair of white gloves [laughs]." Donning white gloves and other symbols of femininity and wealth, these advocates employed embodied devices to make themselves and RCCs appear more credible and innocuous to government actors.

As a result of these efforts, in June 1980, the OCRCC was granted $450,000 over three years (Statement in the legislature, 1980). The amount was significantly less than the OCRCC originally requested, and it came with many strings that would continue to limit the activities of the OCRCC in later years. However, in 1980, it offered much needed financial support for the RCCs in Ontario.

These activist fundraisers, like the advocates in the medical exam room, had to strategically present themselves and their claim to expertise as not only non-threatening, but also crucial to sensitive care and services for rape victims. The funding negotiations, as well as the law reform consultations, training sessions, and forensic exams, were far from the collaborative, equitable meetings between RCC members and institutional actors that government historical records suggest. Instead, these meetings were complicated interactions shaped by inequalities of status and power and contestations over expertise. These complicated relations formed the backdrop to the controversies over standardized medical protocols for evidence collection and the eventual construction of the SAEK.

Designing the Sexual Assault Evidence Kit

In the late 1970s, anti-rape activists and feminist scholars began calling for standardized protocols for forensic evidence collection in rape cases. Despite the value that investigators and forensic scientists placed on forensic evidence and the apparent authority that it had in law, defence lawyers often successfully contested the evidence in court and many judges deemed it inadmissible on the grounds that hospital staff had not collected or stored the evidence properly (Feldberg, 1997; Parnis & Du Mont, 2006). Defence lawyers argued that the quality of the forensic evidence was significantly reduced by long delays in evidence collection in hospital emergency wards and hospital staff's mishandling of the evidence after it was collected (Kinnon, 1981). As defence lawyers challenged medical forensic evidence, they cast a shadow of doubt over what was believed to be the inherent objectivity of medical practice. Without standardized protocols dictating practice, action in emergency wards could be more easily argued to be unreliable. Without a formalized system of coordination between actors collecting and handling evidence (physicians, nurses, police, and forensic scientists), the evidence could be dismissed as being mishandled and contaminated. These legal arguments made the subjectivities of medical practice more visible in the courtroom and challenged the status of forensic evidence as a credible, objective witness of rape. As a result, many rape cases ended in acquittals because of a lack of viable medical evidence (Marshall, 1980). In the late 1970s, anti-rape activists and feminist scholars were becoming increasingly aware of these trends and began arguing for standardization of medical forensic practice in rape cases.

Clark and Lewis's (1977) book on rape exposed Toronto hospitals' reluctance to respond to rape cases, the lack of training for medical examiners, and the lack of uniformity in medical evidence collection procedures. Clark and Lewis asserted that standardized medical protocols would not only improve forensic evidence collection but would also ensure more sensitive treatment of victims. Standardization, in Clark and Lewis's estimation, held the dual promise of creating more objective evidence *and* more sensitive care – an argument that was reiterated by many others.

Joanie Vance (1978) drew a clear connection between standardization and better care for rape victims in her RCC report to the National Department of Health and Wellness. She wrote: "Standardizing protocols for the treatment of victims of sexual assault is necessary at every level ... if the sensitive handling of all victims is to be assured" (3). Many RCC advocates agreed. One advocate I interviewed summarized the fight for standardized protocols by saying that they wanted "to make sure that regardless of the hospital that you went to, you were going to receive the same care, the same procedure was going to be followed ... whether you were in St. Catharines, Welland, or Niagara Falls."

If the forensic exam could be standardized with an evidence collection kit that contained protocols and necessary tools for evidence collection, many advocates hoped that physicians would be forced to give victims better care and collect more credible evidence. One retired RCC advocate explained that their motivation to fight for a standardized evidence collection kit was "to make sure that when we got to court things weren't thrown out ... that you didn't give the defence the opportunity to challenge how the collection was done ... because you had something standardized, it wasn't individual based on who had done the exam." This standardized kit held the potential in their eyes to reduce a physician's biases and prejudices against female sexual assault victims and to remove traces of subjective individual action in medical forensic evidence. To return to Haraway's (1997) term, the kit held the promise of transforming physicians and the evidence they collected into more objective, *modest witnesses* of the injuries and traces of rape on a victim's body. The technology of the kit seemingly promised to introduce and maintain the objectivity that advocates saw as so lacking in medical and legal practice. By eliminating traces of the individual examiner and their biases, activists hoped that forensic evidence would appear more objective in court and would lend credibility to the victims who seemingly had none in the courtroom. With a standardized kit, many advocates anticipated that forensic evidence would be more likely to be deemed admissible and rape conviction rates would rise.

These arguments to standardize Canadian medical practice around rape coincided with developments of standardized protocols in the United States. In 1974, Dr Fahrney, an American physician, proposed a standardized "sex assault kit" (Fahrney, 1974, 340) for emergency wards and vowed that it would "speed up the proper examination of the patient, the adequate collection of the medical evidence, the labeling of specimens, and [be] a method by which the data can be submitted rapidly to law enforcement officials" (340). This kit, he claimed, would be the solution to "one of the most difficult problems for the busy emergency physician to handle" (340), which were in his estimation, rape cases.

In the late 1970s, some Ontario RCCs began working with hospitals and police in their communities to develop standardized medical protocols for rape cases. Not all activists in the movement supported this work and some challenged the merits of collaborating with institutions in this way (Rebick, 2005; Zook, 1980). Among those who did work with hospitals and police, developing institutional protocols meant for some "having to work in paradigms not of our choosing," as one retired advocate put it. When they negotiated with medical and legal professionals, advocates had to accept the value of the medical exam, at least enough to argue for its improvement. According to the advocates whom I interviewed, some readily accepted the medical exam and forensic evidence as important tools that lent credibility to women's experiences of rape. Others did not. One rape crisis staff member explained,

> You could sort of see how things were going ... As long as the justice system and the media were going to refer to this as evidence, as long as the police felt that this was necessary in order to bring a case to trial ... then we wanted women to have the best care possible. But we didn't agree with

the basic premise [of the forensic medical exam] to begin with. I think that was often a position that the centre was in, we wanted to advocate on behalf of women, but the basic premise that we were working from was wrong.

Despite these internal struggles in RCCs, some advocates turned their attention to the task of standardizing forensic medical evidence collection.

A Project of Inscription

In November 1978, the Niagara Region initiated the Cooperative Care Project, which aimed to test the value of standardized protocols for medical forensic evidence collection in rape cases. Driving the Cooperative Care Project was the Committee Against Rape and Sexual Assault (CARSA), the rape crisis centre in the Niagara region. Unlike many other RCCs at the time, CARSA's board of directors included not only RCC staff members, but also nurses and police officers. This mix of actors set the stage for discussions of procedural reforms in medical and legal institutions.

In the months preceding the project's unveiling, experts gathered to design a standardized kit for evidence collection, which was to be piloted in the seven Niagara region hospitals. Forensic scientists from the Centre of Forensic Sciences in Toronto flew to St Catharines to hold a series of design consultations with Niagara police, physicians, and RCC staff (CARSA, 1980). There are few historical records of the discussions that occurred in these design consultations.[10] Much of this history has been lost with the passage of time. However, the political and institutional contexts in which all of these actors were working paint a picture of the complicated dynamics that might have occurred at the design table. Anti-rape activists' ongoing struggles to assert their expertise and credibility in law reform consultations, training sessions, funding negotiations, and medical exam rooms likely appeared in aspects of the design consultations as well. Although some CARSA advocates were invited to the consultations, the power relations in the room likely shaped their ability to significantly influence the kit's design. Just like the advocates who were navigating tense relations with physicians in hospital exam rooms and the activists who were fighting for RCCs' security in funding negotiations, the advocates at the design table likely had to carefully perform their expertise to the police, physicians, and

forensic scientists in the room. The activists' contributions to the discussions may have been influenced by the tenuous relationship that
RCCs were building with government funders, their increasing vulnerability due to lack of resources, and the ongoing marginalization
of their expertise and knowledge in medical practice. Their ability to
advocate for a kit design that reflected victims' needs, and not just the
needs of an evidentiary-based legal system, could have been hampered
by the differences in professional status, credentialized training, and
institutionalized expertise between many of the RCC advocates on
one side of the design table and the scientists, lawyers, and police on
the other. Leaning on the credibility and authority of law, science, and
medicine, the physicians' assertions about rape treatment, the police
officers' claims about evidentiary requirements in rape law, and the scientists' statements about proper evidence handling and preservation
likely overshadowed many advocates' statements about rape trauma
and victim's needs.

Designing a standardized kit meant visualizing a new forensic tool
that would standardize medical practice and produce more reliable,
credible testimonies of rape. This new tool had to *translate* the complexity of victims' experiences of rape into evidence that the legal
system would deem relevant in a rape trial. Because law and legal
practice demanded corroborative evidence of injuries and physical
traces of rape, the kit was built to collect this kind of evidence. The
designers assembled a cardboard box that was filled with bags for the
victim's clothing, swabs for trace evidence, vials for urine and blood
samples, combs for hair samples, and envelopes for fingernail scrapings. They developed standardized instructions for physicians to follow when collecting evidence, and forms for mapping victims' injuries
and documenting their emotional state. They envisioned the kit being
shipped from physicians' exam rooms to police stations to forensic
labs, and imagined forensic scientists and technicians analysing and
transforming its contents into forensic evidence that could be used as
objective proof of rape in court. In building this standardized kit for
medical examiners, the designers had to imagine a boundary object
that could help to coordinate practices in law, medicine, and science.
Most significantly, they imagined a technology that could be used to
witness rape trauma in ways that reflected and supported legal understandings of rape.

Legal and medical histories of viewing the victim's body as a site of
proof of rape and discounting and doubting her account of the assault

were reflected in the kit's design. With swabs, envelopes, vials, and combs, the designers assembled a technology that focused on witnessing *visible* evidence of rape on a victim's body – physical injuries, hairs, bodily fluid stains, and observable emotional states. The emotions and traces of rape on the victim's body that the physician could observe were included in the kit's purview. However, the victim's subjective experiences of rape were not. Although the kit's standardized forms included space for documenting some aspects of victims' reports of rape, physicians were responsible for completing these forms, not victims. The forms made no room for the victim to record her subjective experience in her own terms. Instead, her experience of rape was translated into forensic language by the physician and the standardized forms that they used. The forms stipulated the parts of her rape that were deemed to be medically relevant for the exam and legally relevant for court, such as the traces bodily fluids on the victim's body and the victim's visible injuries and emotional state. In a legal context where women's reports of rape were seen as unreliable and not credible, women's subjective experiences of rape had little evidentiary value on their own, and as such, they were made invisible in the new standardized evidence kit.[11] The distrust and dismissal of women's reports of rape and the corresponding reliance on corroborative forensic evidence were *inscribed* in the kit's design. By containing tools for collecting visible traces of rape and forms that largely ignored the victim's subjective experience of rape, the kit embodied and reflected a political context in which women's reports of rape had little credibility. The politics of doubting victims in medicine and law was thus inscribed into the new kit and how it would witness rape trauma in the exam room.

The kit was to be a technology of translation, which would work alongside physicians and nurses to translate rape into forensic evidence. Law (2007) argues that translation, or the attempt to make two things equivalent, always involves a betrayal. By focusing solely on the visible traces of rape and excluding much of the victim's subjective experience of rape, this new kit would betray the complexities of a victim's experience of rape in favour of producing corroborative evidence of visible injuries and bodily fluids on the victim's body. The new kit reflected the politics of a legal system in which victims' experiences of rape had little credibility. But paradoxically, this tool was to serve as an objective technoscientific witness of rape in rape investigations and trials.

Following the design consultations in 1978, Niagara region doctors and nurses used the new standardized kit for eight months, a process that the Cooperative Care Project committee later evaluated for the Provincial Secretariat for Justice. In the provincial consultations on rape that would follow, this Niagara kit and the Cooperative Care Project would become key figures in the struggle for provincial standardized protocols for rape response.

Demarcating Expertise: A Consultation on Rape

The news media were abuzz in Toronto in the late 1970s with reports that rape rates in the city were on the rise. The number of rapes reported to police had increased by 18% from 1973 to 1978, with a 37% jump from 1976 to 1977 (Provincial Secretariat for Justice, 1978). With the increasing visibility of RCCs and the related heightened public consciousness around rape, more women were undoubtedly coming forward to police to report incidences of rape. However, there was a flood of activity among police and government agencies to recast the statistics as inaccurate and misleading. The OPP released a report in 1978 that blamed the rise in rape rates on women's "promiscuous" and "indiscriminate behavior" (as cited in Kinnon, 1981, 71) and claimed that 71% of rapes reported indicated "some indiscretion on the part of the victim" (Chaddock, 1979, 1). This unleashed a wave of anger and protests from anti-rape activists (Women rally, 1979). A Toronto based anti-rape group, Women Against Violence Against Women, organized a rally at City Hall and released a press release that read: "The OPP report [is] dazzling in its illogic ... [It] illustrates most clearly that, despite our protests and despite our various attempts to explain and define women's experience with rape, our police force has much to learn" (Women Against Violence Against Women, n.d., 1). Feminist *Toronto Star* columnist Michelle Landsberg (2011) chimed in early in 1979:

> The OPP report on sex crimes is breathtakingly inept. It libels the victims of rape as "promiscuous" without a shred of evidence ... The OPP's report breathes new life into the old libel. No wonder more rape victims call the Rape Crisis Centres than call the police ... Perhaps it's too much to ask the police to understand the sickened outrage, the sense of betrayal women feel when society conspires in the big lie that women invite, provoke, and enjoy rape. (119)

Several months before the OPP report was released, the Provincial Secretariat for Justice pre-empted the OPP's findings and claimed in media interviews that the apparent rise in rape rates was in fact illusionary because "a large proportion" of reported rapes were in fact unfounded after police investigation (Unknown, 1978b). Media interview briefing notes for the PSJ reminded politicians that when speaking to the media, "it is important that alarmist statements not be allowed to remain unchallenged and that women across Ontario be reassured that their safety is not in great jeopardy" (ibid., 3) and that "we [the PSJ] continue to be very sensitive to the needs of actual rape victims, even though their numbers are small" (Welbourn & Lambert, to D. Sinclair, 1 December 1977, 4). These media statements did little to appease the public and pressure mounted on government, hospitals, and police organizations to address the systemic problems in rape response (Women march, 1977).

Amidst the heightening pressure, the PSJ organized a Consultation on Rape in February 1978. It marked one of the first provincial efforts to develop and define expert knowledge and practice on rape treatment and response. According to some internal PSJ memos, the consultation organizers were determined to keep the consultation small, so as not to "give credence to the idea that the incidence of rape is rapidly increasing" (Welbourn & Lambert, to D. Sinclair, 1 December 1977, 4). Only fifty-three people were invited to the closed consultation, most of whom were government employees, police, lawyers, doctors, nurses, and forensic scientists. Eight RCC advocates from Toronto and a few other large urban areas of the province were on the invitation list (Provincial Secretariat for Justice, 1978). No invitees were identified in the records as victims of rape.

Through a carefully chosen invitation list, the PSJ was assembling an exclusive expert community on rape. This involved a form of "boundary work" (Gieryn, 1983, 781), whereby some people were designated as experts on rape, and many others were not. While some RCC advocates were invited to the consultation, the PSJ was reluctant to open the consultation to more activists in the anti-rape and women's movements, as a letter from Provincial Secretary for Justice George Kerr revealed:

> While we received an enormous number of requests to attend ... particularly from women's groups ... we did not accede to those requests ... because we wished the Consultation to be oriented to action and could foresee that the presence of a very large number of interested citizens carried with it the distinct possibility that Consultation could become a stage for political posturing, or for the achievement of opinions rather than

facts, emotionalism rather than an objective and reasoned approach to a very real problem (Kerr, 1978a, G. Kerr to R. Jeffrey, 16 June 1978).

By drawing a distinction between objective, reasoned professionals and emotional, political citizens, Kerr was demarcating the boundaries of a small, select group of experts on rape. He created those boundaries by labelling others outside of the group as emotional non-experts whose contributions he predicted would not only be irrelevant, but also a threat to the reasoned discussions on rape. Embedded within Kerr's boundary drawing between these two groups, was an emerging definition of expertise on rape as being objective, reasoned, and factual, and never emotional or political. With the PSJ consultation being one of the first provincial government efforts to formalize expertise on rape, the PSJ's exclusion of many activists and victims, and the boundary work that it implied, is particularly significant.

Historical records from the consultation reveal that much of the discussion was geared towards developing a more coordinated, standardized medicolegal response to rape, in which a standardized protocol for evidence collection would be a part (Provincial Secretariat for Justice, 1978; Sinclair, 1978). George Kerr set the tone in his opening address at the consultation, where he outlined the challenges that rape cases presented for law enforcement and stressed the importance of medical evidence of rape. Referring to victims and accused, he said, "It is quite common for the two parties involved to interpret the situation quite differently. Without clear evidence of a physical attack, it may be almost impossible to determine with confidence which interpretation is correct" (Kerr, 1978b).

One panel at the consultation took up Kerr's call to action and discussed medical evidence and victims' treatment in hospital emergency wards (Provincial Secretariat for Justice, 1978). The panel participants spoke at length about physicians' continued reluctance to treat victims, physicians' lack of training in medical forensic evidence collection, and physicians' "judgmental attitudes" (Provincial Secretariat for Justice, 1978, 4). They highlighted the lack of standardized protocols for medical forensic evidence collection as a serious concern. The panel made two recommendations: that "a standardized protocol and procedure for hospital staff be established for victims of rape ... [and] the development and use of a standardized Rape Kit for all hospitals in the province" (ibid., 5). This recommendation was likely the first government-sanctioned call in Canada for a provincial standardized kit for collecting and analysing rape case evidence.

In the consultation's closing remarks, Don Sinclair, the Deputy Provincial Secretary for Justice, applauded the participants' efforts and emphasized the importance of a coordinated response to rape between all experts in the field (Sinclair, 1978). Without coordination and cooperation within this new expert community on rape, he claimed, there would be no progress for rape victims. He proposed a new ethic of engagement for this expert community, stressing the importance of cooperation and the dangers of criticism:

> The need for cooperation has been evidenced throughout the Consultation and to meet that need it is incumbent upon each of us to acknowledge that we are not the only players in the game and that others' roles are just as important as ours. We shall make no progress by criticizing the other players ... We lose our effectiveness and our credibility when, by direct accusation or by implication, we leave the impression that we (whoever "we" are) are doing the "right" things while others fumble or bumble or just don't care. *Nobody has a monopoly of expertise or of good will in this area."* (Sinclair, 1978, 10, emphasis added)

By referring to consultation participants as "players in the game," Sinclair invoked a sporting metaphor to describe rape response, in which nurses, doctors, police, lawyers, and rape crisis workers were all players on the same team. At a time when the PSJ was building a claim to expertise on rape, this was a strategic configuration. His words implied that a new team of experts had emerged from disparate groups of actors. On this new team, critique against fellow players would be discouraged. For those anti-rape activists who, since the early 1970s, had been denouncing medical and legal institutions for failing to take rape seriously, the choice was made clear: either play by the rules of the team or be cast to the sidelines of this new expert community.

Sinclair's claim that "nobody has a monopoly of expertise" levelled the expertise in the room. The years of the anti-rape movement building knowledge on women's experiences of rape were discursively pushed aside. By denying everyone a claim to expertise, Sinclair opened the opportunity for professional groups, who had, up until that point, expressed little interest in the problem of rape, to claim a newfound expertise on rape. Sitting on the cusp of what would be a rise in specialized, professional knowledge around rape and sexual violence in the

1980s and 1990s, Sinclair's remarks laid the foundation for the development of medicolegal expertise on rape.

Assembling Ontario's First Sexual Assault Evidence Kit

Following the consultation, an implementation committee, consisting of ten consultation participants, two of whom were RCC staff, worked to implement recommendations from the consultation (Unknown, ca. 1978a). The committee used the kit that had been designed in the Niagara region to develop a standardized Sexual Assault Evidence Kit for the province of Ontario. The first provincial SAEK was literally assembled by the hands of RCC volunteers and staff. The Niagara RCC received a small contract from the PSJ to prepare and pack 3000 kits for the province. One activist remembered, "I remember sitting with 3,000 boxes and just putting together all these kits, which then would have been distributed to the hospitals in the province." Despite the hesitations that some RCC advocates had with the kit and their complicated and often marginalized role in medical forensic practice, advocates were again drawn into the assembly of the SAEK.[12]

The provincial government introduced the SAEK to Ontario on Friday, 16 January 1981.[13] In a press release announcing its arrival, the minister for justice policy, Gord Walker, applauded the new technology and its potential to transform criminal investigations of rape:

> Standardization of a sexual assault evidence kit is one further step in our efforts to help the victims of crime ... The extent of the consultation among practitioners in the health and justice fields demonstrates the seriousness we attach to finding better ways of helping the victims of sexual assault ... This new kit should shift the responsibility for identifying evidence onto the *experts* in health and forensic science and thus alleviate some of the pressure on the victim, at a time of great emotional distress, to *produce evidence* of a criminal attack ... and provide the police with evidence necessary for a thorough criminal investigation. (Provincial Secretariat for Justice, 1981, 1, emphasis added)

Walker used the development of the SAEK as a political testament to government and medicolegal actors' active interest in addressing victims' needs, despite what had been a documented absence of this interest throughout much of the 1970s (Clark & Lewis, 1977; Donadio & White, 1974; Williams & Williams, 1973). His claim that the kit would "shift responsibility onto the experts" foreshadowed the growth in expert knowledge and practice on sexual violence in the 1980s and 1990s. Most significantly, his statement conveyed the promise of a new technoscientific witness that would coordinate medical and legal practices and produce objective, credible evidence of rape.

The kit's announcement expressed high hopes for the new tool. Walker predicted that the SAEK would be "a significant breakthrough in helping women prove sexual assault" (2). Given the politics that were inscribed in the SAEK – the distrust of women reporting rape and the presumed necessity of corroborative medical evidence – the extent to which the kit would actually prove to be a "significant breakthrough" for victims was debatable and would become the source of much controversy. Over the course of the next three decades of the SAEK's life in Ontario, some would tout the tool as invaluable while others would condemn it as woefully inadequate and inherently harmful.

3 Stabilizing the SAEK: Controversies in Practice, Advocacy, and Expertise

You could hear the gears of specialization grinding, the carving up of victim-populations, the negotiations of turf, the vying for funding, for prestige, for place. Never having heard it before, I did not then identify the hum and buzz as the sound of persons professionalizing.

> Toronto Rape Crisis Centre / Multicultural Women Against Rape (n.d.)

Just a few years after the first Ontario Sexual Assault Evidence Kit was unveiled in 1981, the media branded the new tool as "Ontario's most successful rapist trap" (Crawford, 1984, A13).[1] The SAEK, media reports claimed, had "revolutionized the way evidence [was] collected in sexual assault cases" (A13) and had almost doubled the conviction rate for sexual assault to 70%.[2] Gord Walker, the minister for justice policy at the time, applauded the new kit: "I have files inches thick with letter after letter praising the kit. This is a little success story in the collection of evidence" (A13). The SAEK, so it seemed, had been immediately accepted in medical and legal institutions as a trusted technoscientific actor that "trapped rapists" with its reliable, credible testimony of rape. However, behind these success stories of the SAEK, the new tool was provoking new controversies about medicolegal expertise, rape crisis advocacy, and forensic evidence collection. This chapter reveals how medical, legal, and government actors fought through these controversies to stabilize a medicolegal network in which the SAEK could be accepted as a trusted and credible tool.

SAEKs were distributed to emergency rooms across the province in the early 1980s. By 1983, 82% of the hospitals surveyed by the Ontario Hospital Association had at least one SAEK in their emergency ward

(A report of the OHA survey, 1983). Integrating this new actor into medicolegal practice and developing consensus around its purpose and value proved to be far more difficult than just equipping emergency wards with SAEKs. When the kit was first introduced to Ontario hospitals, many physicians saw it as an unnecessary burden, while government agencies saw it as an essential tool for rape investigations and prosecutions. Some advocates viewed the new kit as a sign of institutional progress, whereas others saw it as an extension of patriarchal institutions. In an effort to develop some consensus around the kit's value as a technoscientific witness of rape, medical, legal, and government actors created new experts, practices, and institutional spaces to support the SAEK and its new role in medical and legal responses to rape. In the face of ongoing controversies, disputes, and uncertainties, they sought to build and stabilize a network of experts and expert practices around the SAEK.

This chapter explores the work of building networks and reveals the efforts to build and stabilize the SAEK's medicolegal network in the 1980s and 1990s. I use the term network as a metaphor for understanding the relations between the SAEK and the hospital staff, police, lawyers, and rape crisis centre (RCC) advocates who worked alongside it. As I discussed in chapter 1, actor-network theorists have used the concept of socio-technical networks to draw attention to how social and technical worlds are co-constituted within relations between human and non-human actors. Here, I examine the complicated relations between actors within the first SAEK's *medicolegal* network, which featured a confluence of the social and the technical, but more specifically, a confluence of the medical and legal.

Feminist scholars have criticized ANT studies for their limited analysis of power relations between actors in socio-technical networks (Fujimura, 1991; Star, 1991; Wajcman, 2000). Feminist scholar Joan Fujimura (1991) outlines her departure from Latour and other ANT scholars when she writes,

> I want to examine the practices, activities, concerns, and trajectories of *all the different participants* – including nonhumans – in scientific work. In contrast to Latour, I am still sociologically interested in understanding why and how some human perspectives win over others in the construction of technologies and truths, why and how some human actors will go along with the will of other actors, and why and how some human actors resist being enrolled. (222)

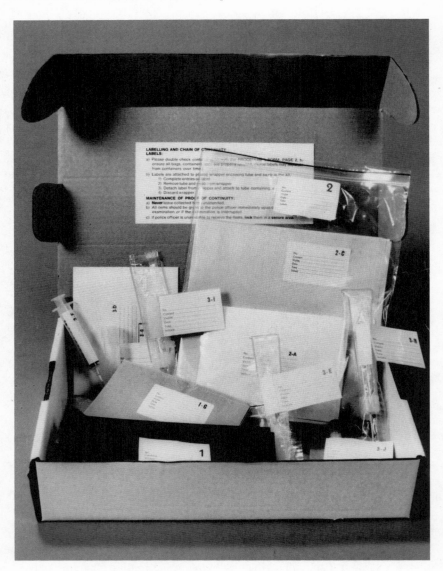

Figure 3.1 Ontario Sexual Assault Evidence Kit, 1984. First appeared in the *Toronto Star*. (Reprinted with reproduction licence from Getty Images)

This chapter traces the work that took place in rape crisis centres, hospitals, and government boardrooms in the 1980s and early 1990s to build and stabilize a coordinated network of relations around the SAEK. Taking up a similar approach to Fujimura's, this chapter reveals the relations between human and non-human actors in the SAEK's medicolegal network, and the power relations and controversies between and within groups of anti-rape activists and medical and legal professionals.

While the SAEK's medicolegal network was being built and stabilized in the 1980s and 1990s, so too was the SAEK. As medical, legal, and government actors were trying to resolve controversies over the SAEK by developing new experts and institutional spaces for its use, the kit was becoming further entrenched in medical and legal practices that rarely met victims' needs. In addition, as the kit gained status as an objective tool, its political origins in legal histories of distrusting victims of rape were becoming less visible. By charting the work that went into building and stabilizing the SAEK's medicolegal network, this chapter also reveals how the kit itself gained stability in medical and legal practice as a reliable, credible technoscientific witness of rape.

The efforts to stabilize the SAEK and its medicolegal network marked an attempt to create a more powerful and coordinated system of experts and expert practices around the SAEK; a system in which victims' bodies were constructed as crime scenes, and medical and legal actors as experts in rape victim advocacy and treatment. In this network, the SAEK served as a boundary object that drew together and coordinated practices in medicine, science, and law. Halfon (2010) argues that stabilized networks feature coordinated action and shared frames of understanding. Stabilized networks, Halfon suggests, are strong and cannot easily be subverted. Stabilizing the SAEK's network required coordinating how physicians, nurses, advocates, and victims used the kit, and the meaning that hospital staff, advocates, police, and lawyers drew from its contents. Building this kind of stability through controversy, changing laws, and shifting practices proved to be no easy task. In the wake of a decade of anti-rape activism in the 1970s where activists were invading, opposing, and challenging medical and legal institutions, the efforts to build stability in the SAEK's network in the 1980s and early 1990s were distinctly political. Not only did these efforts help to position the SAEK as a reliable, credible tool in medical and legal practice, they also drew new boundaries around medical and legal responses to rape that pushed anti-rape activists to the margins of medicolegal practice. The kit helped to facilitate this "boundary work" (Gieryn, 1983, 781) of demarcating experts from non-experts in rape treatment and advocacy. All of this action within the SAEK's network occurred against a backdrop of changes in the anti-rape movement, shifting legal definitions of rape, and ongoing disputes in rape crisis centres, hospitals, and government boardrooms about sexual assault treatment, care, and advocacy.

In the struggle to build stability in the SAEK's network, activity around the SAEK was becoming increasingly bounded and defined; anti-rape activists' political work was being regulated and controlled by government funders, physicians' and nurses' work was being coordinated and defined by the SAEK's forensic script, and new actors were claiming expertise and developing new institutional spaces for the SAEK. All of these changes were essential to building stability, but were met with much resistance from advocates, hospital staff, and law enforcement officials. By highlighting this resistance and the controversies it generated, this chapter reveals the struggles and conflict embedded in the building of the SAEK's medicolegal network.

Shifts in the Anti-Rape Movement

The 1980s was a vibrant and transformative decade for the anti-rape movement. In the context of broader funding cuts to social services (Pierson, Cohen, Bourne, & Masters, 1993), pressures to professionalize rape crisis advocacy with specially trained psychologists and social workers, and increasing government regulations on rape crisis centres' work, the anti-rape movement was changing. The movement was reshaping its politics and identity, while simultaneously growing in strength and visibility. Outside rape crisis centres, some feminist lawyers were actively lobbying for legal reform (Prentice et al., 1996). Inside RCCs, activists were devising new modes of organizing and resisting rape; Take Back the Night marches and rape confrontations were becoming part of anti-rape political tactics (Neigh, 2012). All of this work occurred alongside internal debates in the anti-rape movement about its politics and relationship to medicolegal practice. Describing all these changes, one activist said, "We ourselves were transforming and being transformed."

The radical feminist analysis of rape that had dominated much of the anti-rape movement in the early 1970s was heavily criticized in the late

1970s and 1980s. Many women of colour, Indigenous women, disabled women, and women in poverty challenged white, middle-class, radical feminists' theorizing on rape by arguing that it hinged on a false universalism that assumed that all women experienced sexism similarly (Rowland & Klein, 1996). New feminist literature on rape by Canadian and American women of colour critiqued the then popular radical feminist analysis of rape as a tool of sexism. These scholars and activists argued that rape was motivated not only by sexism, but also by racism, classism, ableism, and homophobia (Davis, 1983; Harris, 1990; Klein, 1992; Monture-Okanee, 1992). Some criticized white anti-rape activists for ignoring and silencing women of colour's experiences of rape (Davis, 1983). As Makeda Silvera said in a published conversation about racism and sisterhood, "I'm sick of some of these white feminists when they talk about rape. It's always from *their* perspective – being knocked down somewhere in a dark alley or a park and being raped. They never mention other kinds of rape, other abuse that women of colour and immigrant women experience" (Bannerji, Brand, Khosla, & Silvera, 1983, p. 8). Several anti-rape activists whom I interviewed recalled the tensions in Canadian RCCs when women of colour tried to encourage their white, middle-class sisters to acknowledge their own class and racial privilege; recognize oppression based on race, class, ability, and sexuality; and integrate this understanding into anti-rape activism. One activist who identifies as a woman of colour remembered, "It was an ever struggling battle to have those discussions that moved from just talk to action." Over time, some changes came to pass in the Ontario Coalition of Rape Crisis Centres (OCRCC), and by 1991 the OCRCC reported that the rape crisis centres in their coalition offered more services that accommodated some of the differences among women (OCRCC, 1991). However, several anti-rape activists I interviewed contended that tensions in the movement were ongoing, as many women of colour, disabled women, lesbian women, and women in poverty continued to fight for representation in the RCCs in which they worked. Through these ongoing struggles, more anti-rape activists incorporated discussions of race, class, ability, sexuality, and other social markers into their analysis of rape and sexual violence, and as a result, the movement's politics were being transformed (Cahill, 2001).

Beyond these internal shifts within the movement, there were many external pressures on RCCs to transform their organizational structures, practices, and relationships with medical and legal institutions. Many of these pressures were imposed through government funding conditions and reopened debates in RCCs about whether centres could, and should, work collaboratively with institutions.

Conflicted Relations and Strategies for Resistance

The Provincial Secretariat for Justice imposed a range of conditions and requirements on Ontario RCCs when it agreed to fund the OCRCC in 1980 (Sinclair, 1980, D. Sinclair to S. Sahli, 26 June 1980).[3] Through these funding requirements, the provincial government could more easily observe and regulate action in RCCs, and put pressure on anti-rape activists to coordinate their advocacy work with medical and legal institutions. The government's impetus to control RCCs was in part a reflection of the neoliberal currents that characterized the political and economic context of the 1980s. Canadian federal politics were influenced by the turn to free trade and privatization that Conservative leaders Margaret Thatcher and Ronald Reagan were cementing in Britain and the United States in the 1980s. When the federal Progressive Conservatives came to power in Canada in 1984, many Canadian social services were stripped of funding and women's groups were cast by some conservative leaders as inherently oppositional and politically threatening (Cohen, 1993). With the widespread turn to neoliberalism in the 1980s, Bumiller (2008) argues, a conflict emerged between the state's commitment to neoliberal values of individualization and privatization and the anti-violence movement's continued call for increased state responsibility to support sexual assault services. This conflict, she contends, diminished as the movement and the anti-violence services that rape crisis centres offered became increasingly regulated by the state. During this period, social service agencies were increasingly controlled by and forced to conform to state-imposed policies and procedures. The Ontario government's funding requirements reflected these broader trends.

In order to receive $150,000 of yearly funding from the Provincial Secretariat for Justice, the OCRCC had to change its collective organizational structure to a hierarchical one, which had to include a head office and a contracted "firm of chartered accountants and legal counsel to assist in conforming to all corporate requirements" (Campling, 1980, C. Campling to G. Walker, 25 March 1980). This condition generated much contention within the OCRCC. Some advocates argued that the collective spirit of the OCRCC would be lost with the newly imposed hierarchical order, while others contended that the OCRCC had no choice but to accept the conditions in order to keep rape crisis centres from closing. The Ontario Coalition adopted the hierarchical structure and many local rape crisis centres followed suit; only a few rape crisis centres retained their collective organizational structure.[4]

The PSJ's funding requirements also stipulated that the OCRCC had to work cooperatively with medical and legal institutions and "explore mutually with the Secretariat the possibility of joining forces with other centres and/or agencies delivering services on a crisis basis [and] continue the Centre's efforts to develop closer liaison with hospitals" (Sinclair, 1980, D. Sinclair to S. Sahli, 26 June 1980). While some RCCs already worked closely with hospital staff and police to reform medicolegal practice, others resisted this imposed condition. Forcing RCCs to develop cooperative and collaborative relationships with hospital staff and police was, for some advocates, a naive and impractical expectation, particularly because, from their perspective, most were loath to acknowledge the value of rape crisis advocates' work (Toronto Rape Crisis Centre / Multicultural Women Against Rape, n.d.). Some anti-rape activists saw advocacy work as deeply oppositional to medical and legal systems and viewed the requirement to "liaise" with institutions as a government effort to depoliticize anti-rape activism and institutionalize anti-rape advocacy.[5]

Debates were also intensifying in the movement about the relationship between RCCs and medical and legal institutions (Neigh, 2012; Pierson, 1993). Some Ontario RCCs grew increasingly frustrated with the inadequacies of medical and legal responses to rape victims and devised alternative, non-medicolegal strategies for responding to male violence. The Toronto Rape Crisis Centre was one of those centres; they offered victims several alternatives to reporting to the police and going to the hospital. Their brochure introducing these options read:

> As rape crisis workers and as women, we all know that the "legal justice" system does not work for us … Taking control and taking action in ways that a woman determines to be best for her are essential parts of women's liberation. Part of our work in Rape Crisis Centres is to take knowledge and information gained from women's experience and put it together so we can offer as many options as possible to assaulted women. (Toronto Rape Crisis Centre, 1986, 1)

The TRCC provided women with three alternatives to the legal system: (a) a personalized letter from the TRCC to a rapist to "let the rapist know that women will not be silent about rape, that what he had done is wrong, and [to] suggest that he seek counseling" (1), (b) a community postering initiative to warn a neighbourhood about a rape and "make the issue of violence against women public" (2), and (c) a rape confrontation to allow a victim, along with a number of TRCC women, to confront her rapist and express how the rape had affected her. The TRCC explained that these alternatives "show we do not need men or institutions to act on our behalf; we know the experience of rape, we tell the truth and we are strong" (2). Not all advocates agreed with these strategies and some reportedly insisted that they were dangerous and misplaced in RCCs (Agencies split over aid, 1984). Others saw them as "wonderfully human" (Lakeman as cited in Neigh, 2012). These debates marked a dynamic period in the movement's history, in which many activists were compelled to comply with government demands to work collaboratively with medical and legal institutions, while others actively tried to resist them.

The changes that government funding conditions imposed on RCCs were part of a larger move to professionalize and credentialize women's organizations, advocacy, and care (Masson, 1998; Morgan, 2002). Medical professional organizations in the 1980s forced many non-professional healthcare providers in the women's health movement, the lay midwifery movement, and alternative medicine to conform to new professional standards (Morgan, 2002). Similarly, in the anti-violence movement,[6] feminist peer advocacy was under attack from regulatory bodies in psychiatry and psychology that sought to regulate and medicalize counselling services for women (Marriner, 2012). These greater demands to professionalize rape crisis advocacy were also solidified in funding agreements in later years, when government agencies placed increasing demands on RCCs to professionalize their advocacy and peer counselling services by hiring psychologists and social workers (Cohen, 1993; Masson, 1998; Ng, 1996). While some centres adopted a more professionalized model of advocacy, others fought to maintain the feminist forms of advocacy that had characterized many RCCs in the 1970s (Marriner, 2012). Reflecting on her centre's refusal to professionalize their services, one advocate I interviewed said, "We never required of each other that you have a degree. You don't have to be a social worker to do good work, to be an effective counselor or advocate ... [Professionalized services] take advocacy out of the community ... and put it in the hands of someone who is a trained professional." The pressures to professionalize rape crisis advocacy prefigured the development of new medical professionals who would specialize in sexual assault treatment and care. Most importantly, these pressures set the stage for RCC advocates' expertise on rape – which was largely rooted in their peer advocacy, feminist politics, and, for some, personal experience – to become increasingly marginalized in medicolegal practice.

The provincial government's efforts to regulate and control RCCs' advocacy were part of a broader move to stabilize a network of medical and legal practices around the new SAEK. Government funding requirements, and the associated pressures to professionalize RCC advocacy, pushed RCCs to mirror the hierarchical organizational structures and professionalized forms of work in medical and legal institutions. In doing so, these regulations sought to silence RCCs' political opposition to medical and legal institutions, and impose more coordination and consensus between them. Although some RCCs actively resisted this imposition, many others in the face of financial instability had little choice but to bend to the government's requirements to lessen their political opposition in favour of "develop[ing] closer liaison(s)" with medical and legal institutions (Sinclair, 1980, D. Sinclair to S. Sahli, 26 June 1980).

Alongside these shifts within RCCs, there were many changes afoot in medical and legal responses to sexual assault. New expertise, experts, and expert spaces were developing around the SAEK, as controversies were developing about its purpose and usefulness. The SAEK's medicolegal network, and the practices and actors within it, were transforming to stabilize the new actor that promised to be a credible technoscientific witness of rape.

Stabilizing the SAEK: A Forensic Script in Practice

There were high hopes for the SAEK when it was first introduced in Ontario in 1981. Government agencies anticipated that it would ease the tensions among law enforcement officers and hospital staff by standardizing forensic evidence collection and increasing the efficiency of forensic exams (Ontario Hospital Association, 1983b; Provincial Secretariat for Justice, 1979a). The SAEK's standardized steps for evidence collection would, the Provincial Secretariat for Justice (1979a) hoped, give forensic evidence greater credibility in rape trials and reduce pressures on physicians to give testimony about medical evidence. These expectations for the SAEK were, however, not so easily realized. Hospital staff did not unquestioningly accept the SAEK as anticipated, nor did the tool resolve the controversies around forensic evidence collection. To the contrary, when the kit was put to use in medicolegal practice, it sparked many debates between government agencies, hospitals, law enforcement organizations, and rape crisis centres that had to be confronted before the kit and the network of practices around it could begin to gain stability.

Technological objects, according to Akrich (1992), are embedded with "scripts" (208) that stipulate rules for action and meanings of use. The SAEK's script was envisioned and brought to life when the SAEK's designers "inscribed" (208) legal histories of distrusting women who reported rape into the new tool. Building on Akrich's notion of scripts, Timmermans and Berg (1997), describe medical protocols as "technoscientific scripts," which, they suggest, define a protocol's "actions, settings, and actors" (275). Technoscientific scripts include not only prescriptions for action in medical protocols, but also the prescriptions for who performs the protocol, where, on whom, and for what purpose. The SAEK was inscribed with a particular type of technoscientific script: a *forensic* script that specified not only steps for forensic evidence collection, but also defined the purpose of these steps, the roles of the actors involved, and the meaning of the evidence collected. It defined the actions that physicians, nurses, and victims were expected to take inside and outside the exam room, and prescribed the meaning that the kit's evidence was supposed to have in medical and legal practice. The kit's forensic script defined physicians' role in the forensic exam room as the expert users of the kit, and defined victims as crime scenes to be examined. The kit's script defined the kit itself as a reliable technology for collecting and mapping traces of rape on women's bodies and for producing credible and objective forensic evidence of rape. Most importantly, the kit's forensic script assigned value to the kit as a technoscientific witness of rape in medical and legal practice.

By prescribing actions and meaning, the kit's forensic script was intended to coordinate medical and legal practices, build consensus around the kit's meaning, and stabilize its medicolegal network. Akrich (1992) suggests that when scripts are "acted out," a "network of technical objects and actors is stabilized" (222). In the SAEK's case, the story was a bit more complex. When the SAEK's forensic script was acted out in the forensic exam room, it standardized and coordinated some of the action in the medicolegal network and imposed some shared understandings of the kit's meaning and value. However, it also sparked new debates about forensic evidence collection, in which the kit and its script were re-evaluated and reformulated. The kit's forensic script was therefore a dynamic set of prescriptions for action and meaning that played a central role in the efforts to stabilize the new tool as credible and necessary in medicolegal practice.

The first provincial SAEK contained many tools and texts for collecting and documenting traces of rape. Hair combs, toothpicks for fingernail scrapings, syringes and tubing for vaginal washes, and swabs for vaginal, anal, and rectal samples were the primary contents of the SAEK, which hospital staff used to collect bodily fluids, hair samples, and debris left on a victim's body. Clothing bags, evidence envelopes, and urine containers were included in the SAEK for storing collected evidence. In addition to these tools, the SAEK also contained texts for standardizing physicians' and nurses' use of the kit. These texts included (1) a consent form for the victim, (2) a medical form for recording a victim's injuries and sexual history, and the details of the assault, (3) a body map for illustrating the victim's injuries, (4) a procedures form that described the SAEK guidelines, (5) a forensic evidence form to document the samples collected, and (6) a procedural booklet for the kit.

The SAEK's texts detailed the procedural aspects of its forensic script by prescribing how the technology was to be used, by whom, where, and under what circumstances. These texts instructed hospital staff to follow the kit's procedures regardless of the specifics of the rape, the victim's emotional state, or the extent to which the staff believed the victim (Procedures form, ca. 1981).[7] One nurse recalled how stringent the SAEK's procedures were: "Once you opened that kit, you were completing every single step in it. You were taking all your patients' clothes, you were plucking head hair. You were doing everything to that patient ... It was horrendous, it really was." Following the SAEK's procedures, so the kit's instructions asserted, would ensure greater efficiency and effectiveness of evidence collection and lend greater credibility to the forensic evidence in court (Provincial Secretariat for Justice, 1979a). The kit's script thus prescribed actions, as well as value and meaning for the kit. Medical and legal actors were expected to understand and accept the SAEK as a technology that would standardize their actions, *and* in so doing produce credible, objective evidence in rape cases.

By prescribing specific actions for hospital staff, the SAEK's script also defined the staff's role in the exam room as active users of the kit who documented and collected traces of rape. Correspondingly, the

Figure 3.2 SAEK Forensic Evidence Form, ca. 1981. (Reprinted with permission from Archives of Ontario)

Ontario

III. FORENSIC EVIDENCE FORM – KIT NUMBER _____

BAG NO. 1 - CLOTHING (itemize)	BAG NO. 2 - BODY EVIDENCE	Done	BAG NO. 3 - VAGINAL AND ANAL CAVITIES	Done
1B	2A Foreign material on body	☐	3A Seminal deposits in pubic hair	☐
1C	2B Seminal stains on skin	☐	3B Combing of pubic hair	☐
1D	2C Scalp hairs	☐	3C Pluck 12 pubic hairs	☐
1E	2D Fingernail scrapings	☐	3D Foreign material	☐
1F	2E Oral swab and smear	☐	3E Vaginal swabs	☐
1G	2F Saliva sample	☐	3F Vaginal aspirate	☐
1H	2G Blood Grouping	☐	3G Slide for motility check motile ☐ non-motile ☐	☐
1I	2H Blood Alcohol/Drug	☐	3H Anal swab and slide	☐
1J			3I Rectal swab and slide	☐
1K			3J Urine sample	☐

Reason, if any of above not done

EXAMINATION
OF (PATIENT'S NAME)

AT (LOCATION - HOSPITAL NAME)   ON (DATE)   (TIME)

OBSERVED BY   CLOTHING AND SPECIMENS RECEIVED FROM:
                                       NAME   TITLE

ASSISTED BY   AT: DATE   TIME   PLACE

SIGNATURE OF M.D. (EXAMINER)   DATE   RECEIVED BY:
                                       POLICE OFFICER'S NAME   RANK

SIGNATURE OF WITNESS   DATE

DISTRIBUTION: COPY 1   FORENSIC LABORATORY
                COPY 2   POLICE OFFICER

kit's forensic script defined the victim's role in the exam room as a largely docile victimized body that SAEK users were expected to act on and around. Mulla (2014) describes how victims in contemporary sexual assault exams often play an active role in guiding forensic examiners to areas of their body where trace evidence is likely to be. While in practice some victims in SAEK exams in the 1980s may have expressed their agency in this way, it was in contrast to the kit's protocol, which configured victims as inactive bodies to be worked upon. The only action that the SAEK protocol specified for the victim in the exam room was signing the consent form,[8] at which point her body became, in the words of some contemporary medical researchers, "a walking crime scene" (Price, Gifford, & Summers, 2010, 549)[9] that physicians and nurses combed for evidence. Her body became implicated in the SAEK's use as it was made the object of others' action: she became an "implicated user" (Clarke and Montini, 1993, 42) of the SAEK. These roles for the victim and the hospital staff were encoded in the SAEK's forensic script and shaped victims' and staff's action in the exam room, and the meaning that the kit had in that space.

The kit's forensic script was geared towards enacting a new actor that would work alongside physicians and nurses to document traces of sexual assault on a victim's body: the so-called scene of the crime. The SAEK's script reduced medical forensic exams to a series of standardized steps. It aimed to standardize physicians' and nurses' empirical assessment of a victim's body and direct hospital staff's attention to particular areas of the "crime scene" where traces of rape where most likely to be found. It instructed medical examiners to map and document what they saw on victims' bodies in standardized ways. By directing examiners' vision and action, the kit became involved in the act of witnessing the trauma and trace evidence on the victim's body. Along with the medical examiners, the kit too became a witness, a technoscientific witness, of the crime scene. Some legislators and policymakers hoped that standardizing hospital staff's evidence collection would strip the resulting evidence of signs of the staff's subjective decision making. This, they anticipated, would increase the likelihood that the evidence would appear objective in the courtroom and be more likely to be deemed admissible (Provincial Secretariat for Justice, 1979a). The hope was that the kit's script would impart agency and credibility on the SAEK itself and the evidence inside it. The kit would be a tool with the agency to remember what was seen on the scene of the crime and the credibility to reliably convey that to the courtroom.

The kit's script, for some, held the promise of giving the SAEK's network stability with a tool that could reliably act as a technoscientific witness of sexual assault.[10]

The SAEK's contents and protocols were inscribed with narrow legal definitions of rape as a physically violent attack involving vaginal penetration and historical legal demands for visible corroborative evidence in rape cases. When the SAEK was put to use in medical and legal practice, its forensic script helped to enact and stabilize these legal demands and meanings of rape. The kit's forensic script thus brought to life the law's narrow understanding of rape.

Reflecting the SAEK's roots in the 1970s, when judges and juries in rape trials were reluctant to convict the accused without corroborative evidence of visible injuries or traces of bodily fluids on the victim's body, the new SAEK's procedures directed physicians and nurses to look for visible traces of rape on the victim's body. They stipulated specific technologies for this task of observing traces of rape in the forensic exam. White and Du Mont (2009) describe how sexual assault is "visualized" (1) in the contemporary forensic exam through "optical technologies" (1), such as cameras for photographing injuries, colposcopes for aiding magnification, and staining agents for identifying damaged tissues. By expanding their definition of technology to include not only highly technical technologies, but also simple tools, it could be said that the SAEK in the 1980s included a range of technologies for visualizing rape: combs for gathering hair; body maps for illustrating violent trauma; and swabs, bags, toothpicks, and syringes for collecting blood, skin, hair, and semen. It also identified other technologies in the exam room to be used in the forensic exam, such as the microscope, which was used for viewing motile sperm. Following the procedures of the SAEK's forensic script, hospital staff used these tools to translate the traces of rape that they saw on a victim's body into evidence that could be contained in the SAEK (White & Du Mont, 2009) and later passed through the hands of scientists and lawyers who would transform it into evidence for the courtroom.

In the forensic exam, physicians visualized rape by producing visual representations of the victim's injuries (White & Du Mont, 2009). The kit's procedures instructed physicians to "look for evidence of violence e.g. marks, bruises, lacerations, scratches, [and] fractures" (Provincial Secretariat for Justice, 1979a, 26) and draw and describe what they observed on a body map. Physicians were told to use the body map to detail the "location, dimensions, tenderness, colour, and estimated

Figure 3.3 Body map from Sexual Assault Evidence Kit, ca. 1981. (Reprinted with permission from Archives of Ontario)

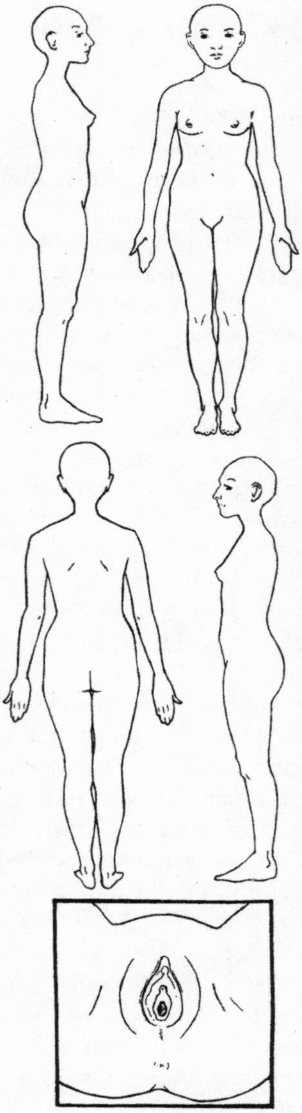

GENERAL EXAMINATION: *(Done by Physician)*
Respect need for privacy and appropriate draping.
Note any trauma on following chart and illustrate on diagrams.

	Bruising/Lacerations	Fractures/Other
Skull		
Face		
Mouth		
Trunk Front		
Trunk Back		
Trunk Right		
Trunk Left		
Upper Right extremity		
Upper Left extremity		
Lower Right extremity		
Lower Left extremity		

GENITAL AND ANAL EXAMINATION — BAG #3

Do Forensic Procedures 3-A to 3-C

Stage of Development _____

Labia Majora _____

Labia Minora _____

Posterior Forchette _____

Introitus _____

Hymen _____

Do Forensic Procedures 3-D to 3-G
Do cervical swab and smear for gonorrhea-Hospital use only

Vagina _____

Cervix _____

Do Forensic Procedures 3-H to 3-J

Uterine Corpus _____

Adnexa _____

Anus _____

Rectum _____

Male Genitalia _____

age of the lesion" (26). These instructions for the body map reflected legal understandings of rape as physically violent and injurious to the victim, as well as the common evidentiary requirement of force in rape cases in the 1970s (Du Mont, Miller, & Myhr, 2003; Parnis & Du Mont, 1999).

Inscribed on the body map itself was the legal definition of rape as forced vaginal penetration that was used in the 1970s and early 1980s. The body map in the first SAEK in 1981 featured a simple line drawing of a female body, which was young, thin, abled bodied, and devoid of any distinctly racialized features (General examination, ca. 1981). Embedded in the body map was an understanding of rape as a crime that could only be committed against female bodies, and an assumption that "real rape victims" (Estrich, 1986, 1088)[11] – or victims that police and the prosecution took seriously – were young, thin females. In addition to the line drawing of a full body, the map also included a close-up drawing of a vaginal opening. The body map constrained physicians' and nurses' visual depictions to the injuries that were caused by forced vaginal intercourse. Rape could only be visualized on the body map as marks on or around the vagina or as less specified marks on the rest of the body. Other physical injuries elsewhere on the body, oral or anal forms of rape, and less-visible emotional and psychological marks left on the victim were obscured or unseen on the body map. By limiting visual depictions of rape, the body map contributed to enacting a narrow medicolegal definition of rape and its victims.

In addition to assessing injuries, the SAEK exam also focused on gathering evidence that could help forensic scientists identify the perpetrator of the sexual assault (Forensic evidence form, ca. 1981). This focus on collecting identifying evidence reflected dominant understandings of rape in the 1970s and early 1980s as being a relatively rare crime that was committed by men whom the victim did not know. It also reflected the then legal definition of rape as an act that could only be committed by a man who was not married to the victim, and the common evidentiary requirement in the early 1970s that corroborative evidence explicitly identify the perpetrator. The assumption that the rapist was a stranger was thus inscribed in the SAEK's material design. The kit's script guided physicians and attending nurses to collect blood, skin, hair, and semen from the victim's body with the vials, slides, envelopes, and swabs in the kit. All of this evidence would be analysed later by scientists and technicians in the forensic laboratory and transformed into visual evidence that could help identify the perpetrator. Lawyers would then interpret the evidence and integrate it into their legal arguments about the validity and truth of the victim's report of rape. Alongside all these other actors, the kit's evidence would be made to "act" in the courtroom as a testament to the crime and the perpetrator's identity. The evidence did not act alone. Its agency was not inherent or independently derived, but instead emerged from its relations with other actors in the medicolegal network.

In the 1980s, forensic identification methods were rooted in a common belief in forensic science that "it is almost impossible to commit a crime without leaving any physical evidence" (Krishnan, 1978, 12) and that physical evidence contained "clues to the circumstances of the crime and the identity of the offender" (12). Individuality, it was assumed, was reflected in the body, and, therefore, individuality could be seen in bodily traces (Cole, 2001). Blood, semen, and hair analyses were the most common forensic identification methods used in rape cases in the early 1980s (Krishnan, 1978). These analyses were predominantly comparative and involved matching samples obtained in the SAEK to those collected from the suspect or from the site where the rape occurred. Through comparison, identity was determined from visible similarities in hair types and/or matching blood types and groupings within blood and semen samples. These techniques could not identify an individual, but instead identified a group of individuals who shared similar blood or hair types.[12] In the 1980s, the kit's forensic script was designed to serve the needs of hair comparison, as well as blood typing and grouping.

The kit's forensic script aimed to create not only a technoscientific witness that worked alongside hospital staff in exam rooms, but also one that could work in courtrooms. In rape trials, the SAEK could act as a witness of sorts, offering testimony on the traces that were witnessed on the victim's body. The kit's script gave the technology meaning and value in the courtroom as a reliable source of medical and scientific facts about a rape. In a context in which victims were deemed to be unreliable and untrustworthy witnesses, the SAEK's forensic script was intended to transform the SAEK into a credible witness that offered objective testimony on rape. This new witness would "not [be] subject to lapse of memory, confusion, and perjury, as are human witnesses" (Krishnan, 1978, 11), as the victim or the medical examiners were expected to be. Instead, this new technoscientific witness could be relied upon to objectively capture traces of rape so they could be further examined in court.

Scientific evidence gains credibility and status as fact in the court-room through visualization (Jasanoff, 1998). The SAEK gave its wit-ness testimony through its visual depictions of rape evidence, and in so doing, it allowed lawyers, judges, and jurors to "virtual[ly] witness" (Shapin & Schaffer, 1985, 60) rape. Shapin and Schaffer describe vir-tual witnessing as a "literary technology" (60) that experimental scien-tists in seventeenth century used to "produc[e] in the reader's mind an image of the experimental scene in a way that obviates the necessity for either direct witness or replication" (60). The SAEK's testimony worked in a similar way: it produced images of rape in lawyers', judges', and jurists' minds, which removed the necessity for directly witnessing the crime. These actors could imagine the rape through the visible evidence captured in the SAEK. The kit's forensic script was designed to endow the tool with status as a reliable witness that could provide objective visual depictions of rape.

In addition to its work in shaping action and the meaning of forensic evidence within medical and legal spaces, the SAEK's forensic script worked to define action outside of them as well. By defining the victim's body as a crime scene in the exam room, the forensic script positioned the victim as an untrustworthy medicolegal outsider who had the potential to destroy the valuable physical evidence at the scene of the crime. Mulla's (2014) contemporary ethnographic research in sexual-assault exam rooms in the United States reveals the expectations and pressures on victims to protect the forensic DNA evidence on their bodies and facilitate its successful recovery. She argues that victims are expected to demonstrate "corporeal discipline" (40) to preserve the forensic evidence on their bodies. These expectations on victims were in existence long before DNA testing was used in criminal investiga-tions. In the early 1980s, the SAEK's script stipulated specific actions that victims were expected to take in order to ensure that the evidence on their body was protected and preserved. Many of the texts, informa-tion booklets, and pamphlets on the SAEK outlined strict instructions for victims on how to properly preserve bodily evidence. The Provin-cial Secretariat for Justice (1979b) handbook for victims gave the fol-lowing directives: "DO NOT BATHE, SHOWER, OR DOUCHE. These actions can destroy evidence" (1). The expectations on victims to dem-onstrate corporeal discipline to protect the evidence on their body were thus embedded in the first SAEK's forensic script.

By laying out these expectations on victims, the kit's forensic script defined the ideal implicated user of the kit. The *ideal* implicated user was a victim who knew and followed the script to properly preserve the evidence that the kit required. She was a victim who appeared to work cooperatively with the SAEK. To fit this mould of the ideal implicated user, a victim needed to have access to the SAEK instructions, the capacity and desire to read the dense English instructions, and the ability and willingness to follow them immediately after being raped. For many victims, the SAEK's instructions were likely difficult to remember and to follow in the face of trauma and the long emergency room delays. Many RCCs attempted to make these instructions more accessible to victims. In doing so, RCCs became directly involved in efforts to stabilize the SAEK and its medicolegal network.

Stabilizing through Conflict: Rape Crisis Advocates and the SAEK's Forensic Script

Although many anti-rape activists had advocated for the SAEK's development in the 1970s and some had participated in its early design, there was little consensus over its use. One advocate described this by saying, "We were never big believers in the evidence kit." According to her, many advocates saw the kit as a tool that gathered useless forensic evidence, which could easily be explained away and defended against in court as traces of consensual sexual activity. Another advocate remembered the debates around the SAEK: "There was a lot of tension around what the kit did beyond prove that there may have been some kind of activity that might be labeled as sexual activity ... There was a great deal of conversation [and] fretting."

Some advocates viewed the SAEK as a useful tool for victims, while others saw it as inherently harmful. Reflecting on these tensions, one advocate asserted that the SAEK in the 1980s "lent some amount of credibility" to women's rape reports and "gave some women comfort," while another argued that the kit symbolized "the medical system becoming ... a gate keeper to the legal system," and falsely represented rape "as either a medical issue or a legal issue, as opposed to a social issue." Another advocate contended that the early kit perpetuated the disbelief in women's testimonies of rape in the legal system, and as such "the kit couldn't be understood outside of the context of patriarchy." Despite conflicting views about the kit, many of the RCC advocates that I interviewed recalled that tensions about what the kit did and what it symbolized were weighed against the knowledge that for some victims (often very few), the SAEK provided the necessary evidentiary proof to convict perpetrators of rape. The kit was, according to one advocate, "a necessary evil."

Despite their uncertainties about the SAEK, many RCCs disseminated information about the kit and its script, and in so doing gave the kit and its network greater stability. RCCs produced brochures for victims of rape that detailed instructions on preserving bodily evidence and volunteer training manuals that specified how to convey information about evidence preservation to victims.[13] Many of these brochures and training manuals reiterated government and medical descriptions on how to preserve evidence and the critical importance of doing so. The doubts that many advocates had about the SAEK's value were erased in these descriptions of the kit's script and its inherent value. RCCs' efforts to cast the SAEK in this way were perhaps not surprising given the fact that their provincial funding largely hinged on their cooperative relations with medical and legal institutions. Regardless of the reason, by publicizing the expectations that the SAEK's forensic script had for victims, RCCs contributed to situating the kit as a central component of the medical and legal rape response. While the SAEK's forensic script may have gained some stability through these texts, in other areas of medical and legal practice, its stability was being challenged and disrupted.

(De)stabilizing the SAEK: A Forensic Script in Question

Despite government agencies' hopes that the SAEK would immediately standardize forensic evidence collection and ease tensions among hospital staff, law enforcement officials, and victims, its integration and acceptance in medical and legal practice was not immediate. Instead, medical controversies and legal reforms threw parts of the SAEK's forensic script into question. The kit's forensic script was challenged and re-evaluated as questions were being raised about rape law in Canada and the appropriate users of the kit. Timmermans and Berg (1997) have illustrated how medical actors often challenge medical protocols and "tinker" (291) with them in practice. They argue that this work can have stabilizing effects for the protocol itself. The debates and discussions that sparked questions about the kit's forensic script ranged in scope from broad sweeping dialogues about Canadian legal reform to more localized controversies over who should use the kit. These controversies destabilized the tool enough to give medical actors, advocates, and government agencies an opening to re-evaluate the kit and consider tinkering with its script. They also provided the opportunity for these actors to reassert the kit's importance in rape cases. At the same time that the kit's forensic script was being questioned, its position in medical and legal practice was becoming more secure and the network of practices around it was gaining stability.

Tinkering Scripts with Rape Law Reform

In 1982, after more than six years of lobbying by anti-rape and feminist lawyers and activists for legal reform, sections of the Criminal Code of Canada relating to rape were redrafted. When the reforms came into effect in early 1983, they eliminated the Criminal Code's narrow definition of rape as forced vaginal penetration and created a new section for sexual assault. Mirroring the structure for physical assault offences, sexual assault was defined as a three-tiered offence with graded levels of violence and penalty.[14] This reform gave sexual assault a more expansive definition that included other forms of non-vaginal rape and sexual assault, as well as sexual assault against men, and eliminated the immunity from rape charges that husbands had in the previous Criminal Code. In addition to the changes in sexual assault law, there were several procedural reforms regarding the evidentiary requirements for sexual assault. Most important to the SAEK, the 1976 reforms on corroborative evidence were reinforced with a clearer mandate in 1982, which eliminated the need for corroborative evidence for a sexual assault conviction with the words "no corroboration is required for a conviction and the judge shall not instruct the jury that it is unsafe to find the accused guilty in the absence of corroboration" (Bill C-127, 1982, s. 246.4).

When these reforms were being drafted in Bill C-127, many RCC advocates hoped that the new reform would eliminate the sexist biases previously embedded in rape law and increase the reporting of and conviction rates for sexual offences (Parnis & Du Mont, 1999). Michelle Landsberg (2011) expressed this optimism in her *Toronto Star* column in 1981 where she described the anticipated reform: "The new terminology happily washes away accumulated layers of sexism currently built right into the laws: that only a woman can be raped, only by men, and only when the penis penetrates the vagina" (125). Despite the confidence with which many feminists and advocates welcomed the reform, Landsberg reported that some were doubtful that the reforms would radically shift the legal response to sexual violence and mourned the loss of the term rape, which had animated the anti-rape movement in the previous decade.

Bill C-127 had significance for the SAEK. By revising parts of the law that had been inscribed in its design, the reforms had the potential to call into question much of the SAEK's script. The more expansive legal definition of sexual assault could have led some to argue that the SAEK's purpose of capturing visual evidence of forced vaginal penetration was unnecessarily narrow. Similarly, the elimination of husbands' immunity to rape charges could have led others to argue that the SAEK's purpose of collecting identifying evidence of an offender was based on the outmoded assumption that strangers committed most rapes. By redrafting the legal basis upon which much of the SAEK was built, Bill C-127 had the potential to destabilize the SAEK and the practices around it. This potential was, however, only partially realized.

In late 1982, just after Bill C-127 had been introduced in parliament, the Provincial Secretariat for Justice began designing and distributing a new SAEK to Ontario hospitals (Cornish, 1982, R. Cornish to Emergency Room Physicians, Nurses, Police, Crown Attorneys, 10 December 1982). The redesigned SAEK reflected relatively minimal changes, which included two new forms (a French consent form and a new medical history form) and eliminated one procedure of scraping the victim's fingernails. The only evidence that the SAEK's procedures had been redesigned to reflect the pending legal reforms was its inclusion of instructions for gathering forensic evidence from male victims of sexual assault. The SAEK's body map, which had previously featured a distinctly female body, was redrawn to feature a seemingly unsexed body with a small jaw, similarly sized hips and shoulders, muscular legs, and short hair. Below the close up diagram of the vaginal area, a diagram of the penile region was added. With these changes, the SAEK's forensic script was tinkered with to accommodate the SAEK's new purpose of collecting forensic evidence of sexual assault from women, men, girls, and boys.

These relatively small changes to the SAEK did not reflect the reform's elimination of the corroborative evidence requirement, as Feldberg (1997) and Parnis and Du Mont (1999) have noted. The SAEK still maintained its purpose of collecting evidence of bodily harm that could be used to substantiate or refute a victim's experience of rape and/or sexual assault. The SAEK retained its emphasis on collecting evidence that identified the perpetrator, despite what many expected to be a significant increase in cases involving known perpetrators. While the SAEK's forensic script was only slightly altered by the 1983 legal reforms, other parts of the script were fervently contested in the years that followed.

Challenging Scripts with Users' Discontent

When the SAEK was first introduced to Ontario, physicians were responsible for conducting SAEK exams (Ontario Hospital Association, 1983a).[15] As primary users of the SAEK, physicians were granted a new expert status over medical forensic evidence collection and medical care for rape and sexual assault victims. Nurses commonly worked as physicians' assistants and explained the SAEK procedures to the victim, collected the victim's clothes, and sometimes, with the physician's permission, collected some forensic samples (Procedures form, ca. 1981).[16] Physician-nurse teams were common in SAEK exams (Du Mont & Parnis, 2003); however, the SAEK's script embedded in the SAEK procedure forms and training videos clearly stipulated the physician's authority over the exam and their obligation to conduct SAEK exams (Procedures form, ca. 1981; Taking care, 1990).

Not all physicians welcomed their new responsibility. While government agencies hoped that the SAEK would increase physicians' willingness to conduct forensic exams by simplifying the process, it did not prove to do so in practice (A report of the OHA survey, 1983). Some physicians felt uneasy about their new dual role as medical care providers and SAEK forensic evidence collectors. One doctor described the feeling common among physicians at the time: "I think anytime we are asked to move outside of a primarily medical role, there is a certain discomfort." This clash between medicine and law played out in physicians' active resistance to the kit's forensic script, which designated them as the kit's primary users.

Thirty-five per cent of hospitals surveyed by the Ontario Hospital Association in 1983 reported problems and frustrations with the new SAEK (A report of the OHA survey, 1983). Most notably, physicians reported frustration about the significant amount of time that the kit imposed on forensic exams. One nurse remembered the frustration among many hospital staff and said, "Everyone would do this big yawn if a sexual assault case came in ... They would say 'Oh god, who's ready to be in one room for five hours?'" These concerns were echoed in medical journals, where some physicians complained about the usability and design of the new SAEK. Debates ensued about the kit's practicality. Physicians Herbert and Whynot (1985) denounced the SAEK in a Canadian medical journal: "It has been our experience that the Ontario kit, like most others ... makes a simple examination unnecessarily complicated, time consuming, and expensive ... One can take all appropriate evidence with a few swabs, test tubes, and slides – all available in any office or emergency department" (1453). Another physician, Len Hargot (1985), who was an avid supporter of the SAEK, penned a response that read, "My fear in having a less defined protocol is that we would be taking a step backwards ... The use of the kit is meant to simplify the procedure, not to complicate it" (1453). However, not all physicians shared Hargot's perspective.

Widespread resistance to the SAEK started early amongst physicians. In 1981, the Ontario Medical Association (OMA) announced that physicians would begin charging $150 for every female SAEK exam and $125 for male forensic exams (Montgomery, 1981). After the Ministry of the Attorney General refused to pay the fee, the OMA threatened that physicians would begin charging the "authorities requesting the service" (1), referring to individual police detachments. By 1983, the OMA and the ministry reached a settlement that would grant physicians $185 for each SAEK exam, to be paid by the Ontario Health Insurance Plan (Ontario Hospital Association, 1983a). Despite having secured payment for SAEK exams, many physicians continued to resist their new responsibility for the SAEK. Physicians often prioritized other medical emergencies, and as a result, victims commonly faced long delays in Ontario hospital emergency wards (Women's College Hospital, 1983).

One victim's wait in a Toronto emergency ward in 1982 made front-page headlines. After waiting several hours in Humber Memorial Hospital's emergency ward, the victim was refused treatment and forced to go to Women's College Hospital for the forensic exam. When the victim reported her story to the media, the physicians at Humber were openly criticized for their failure to provide care and forensic services. RCC advocates criticized the systemic delays for victims in emergency wards across Ontario (Stead, 1982). In an effort to regain public confidence, the minister of health publically stated that he would be "writing to every hospital in the province to ensure that never again will a rape victim be turned away untreated" (ibid., 1). His letter read:

Recent disturbing events surrounding the care and treatment of rape victims in the emergency department of some hospitals is causing some concern. I would like to stress that the *first priority* is to ensure that appropriate medical treatment is provided and that the necessary forensic process is carried out with the *utmost urgency* and sympathy. The rape evidence kits have been designed to provide a standardized means of collecting forensic information and evidence. Although I appreciate that it takes time to complete the requirements within the kit, this aspect should not be allowed to delay or alter the provision of adequate care and attention given to the patient ... It is the duty of everyone involved to treat these unfortunate victims with utmost sensitivity. (Ontario Ministry of Health, 1982, 1, emphasis added)

Despite the government's efforts, public relations battles between government parties were aired in the media, and Ontario Liberal and NDP leaders challenged the Conservative government for being disinterested in victims and their needs (Stead, 1982). As a result of the mounting public pressure, the Ministry of Health proposed to move the SAEK out of Toronto emergency wards and into a regional sexual assault treatment centre in downtown Toronto (Meeting notes, 1983).[17] The centre was to be staffed by specially trained physicians and nurses who would be responsible for treating all victims in the Toronto area (Women's College Hospital Public Relations Department, 1982).

Imagining a new institutional home for the SAEK involved re-evaluating and reformulating the SAEK's forensic script, most particularly whom it assigned as the primary user of the SAEK. Introducing a sexual assault treatment centre to Toronto was intended to relieve emergency room physicians of their responsibility for administering the SAEK and create a new set of expert users for the SAEK: designated physicians and nurses trained in forensic evidence collection in sexual assault cases. Creating an expert space for sexual assault evidence collection in Toronto marked the Ministry of Health's bid to rebuild and stabilize the SAEK's network in the wake of controversy.

Stabilizing with New Expert Spaces

Supporters for the proposed regional Sexual Assault Care and Treatment Centres (SACTCs) described the centres as the only answer to physicians' discontent with the SAEK and to emergency room delays for victims (Meeting notes, 1983; Women's College Hospital, 1983). Women's College Hospital pitched the project as follows:

Although revisions to the Sexual Assault Kit have made it easier to use and procedures have been somewhat standardized, there is still a feeling that police cases of this nature are not truly a part of emergency medicine. Feelings cannot be legislated and problems will continue to occur unless centres providing necessary care are doing so. (4)

The Ontario government framed the proposed SACTCs as having even broader appeal. In the context of provincial cuts to health care and corresponding pressures to enhance efficiency and centralize health care services (Murray, Jick, & Bradshaw, 1984), SACTCs emerged as a viable cost-saving measure. Centralizing health care services for sexual assault and eliminating the delays the SAEK caused in emergency wards with SACTCs was cast by the government as a necessary step for reducing the health care costs of sexual assault treatment (Ontario Ministry of Health, ca. 1984).

Many of these arguments for SACTCs were outlined in a municipal report written by the Toronto Task Force on Public Violence Against Women and Children in 1983, which recommended that sexual assault treatment be centralized in five centres in Toronto. The task force had been initiated in response to a series of rape-murders in Toronto in 1982 and had been charged with the task of investigating the criminal justice system's response to public violence. The task force consisted of six working groups with over eighty volunteers, some of whom were advocates from the Toronto Rape Crisis Centre. In its final report, the task force argued that the five new SACTCs in Toronto should be "staffed by a voluntary roster of physicians trained in the management of the rape trauma syndrome and in the use of the rape kit" (Task Force on Public Violence Against Women and Children, 1983, 14). The creation of a new institutional space for the SAEK was tied to creating and enrolling new experts on forensic evidence collection and sexual assault counselling and medical care. It was to be a new expert space for sexual assault care.

On 12 April 1984, the Sexual Assault Care Centre (SACC) at Women's College Hospital opened its doors in Toronto (Lipovenko, 1984). After what had been some initial concerns about funding and staffing, Women's College Hospital found room for a specialized centre, staffed by voluntary nurses and physicians, that included an examination room for SAEK exams and a seating area for medical consultations (Executive Committee Minutes, 1983; Women's College Hospital, 1983). The centre's mandate, according its coordinator Dianne Nannarone, was,

Two fold ... to present an approach to comprehensive *management* which is responsive to the emotional, physical, and medical/legal needs of victims of sexual assault, and to present a *'networking' model* for nurses and physicians and other health care professionals integrating medical and emotional treatment with the criminal justice system as part of the therapeutic process. (Unknown, 1986, 1, emphasis added)

By housing a network of medical professionals, the SACC was a new physical location where developing medicolegal expertise on sexual assault was established and supported. While only a decade earlier, RCCs in the 1970s were commonly recognized as one of the only sites of expertise and experience with rape response, the new SACC was to be a site of institutionalized and professionalized expertise on sexual assault. By seeking to "manage" sexual assault as an "emotional, physical, and medical/legal" (ibid., 1) problem, practices in the SACC enacted meanings of sexual assault as an individual trauma and sexual assault treatment as a medicolegal intervention (Doe, 2003, 2012). These

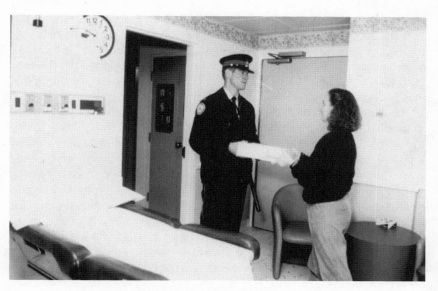

Figure 3.4 Sexual Assault Care Centre, Women's College Hospital, ca. 1984. (Reprinted with permission from the Miss Margaret Robins Archives of Women's College Hospital)

conceptions of sexual assault were a significant deviation from those in RCCs, where sexual assault was more commonly described as a social problem that demanded widespread cultural change. In the SACC, sexual assault treatment was tied to professionalized experts who specialized in forensic exams. This new expert space expanded the SAEK's network by creating and enrolling new experts in the kit's work, and in so doing, wove the tool more tightly into medicolegal practice.

The development of the centre met with resistance. Some anti-rape advocates publicly voiced concerns about the SACC: "I don't want to see mini rape crisis centres set up at hospitals," Deb Parent at the Toronto Rape Crisis Centre said to reporters when the new SACC was announced (as cited in Lipovenko, 1984, M5). Parent questioned why hospitals would want to duplicate RCC services and voiced concern that victim advocacy and counselling would be sidelined to the SACC's medical and legal priorities. With the SACC's specialized experts on the SAEK, she predicted, other physicians and hospital staff would become less willing and able to provide treatment for victims of sexual assault. Despite these critiques, SACTCs began opening in other hospitals across the province not long after the SACC opened at Women's College. By 1987, there were fifteen SACTCs in Ontario (Brodie, 1987) and by 2000 there were twenty-seven (Saltmarche & Cherrie, 2000). In 1993, a provincial network of SACTCs formed and in 1998, many centres' mandates expanded to include victims of domestic violence (KPMG Consulting, 1999).[18] The new institutional home for the SAEK was becoming a fixture in many hospitals across the province.

The media heralded Ontario SACTCs as a significant advancement in medical treatment and care for sexual assault victims (Baker, 1983; Lipovenko, 1984). Many of the RCC advocates that I interviewed reported that SACTCs provided a welcomed alternative to emergency wards for many victims. Nurses expressed similar sentiments. One in particular stressed that the trained SACTC staff and separate location in the hospital created "a more caring and supportive environment" for victims. However, there were undoubtedly many victims who did not directly benefit from the SACTCs. Victims who did not wish to go to the hospital after being sexually assaulted or who could not access the SACTC likely did not feel the widely celebrated benefits of the SACTCs. To access SACTC services, and therefore the SAEK exam, victims had to be willing and able to get to a hospital immediately after being sexually assaulted; they also had to have the transportation necessary to get there, the necessary health care coverage to obtain treatment, the ability to speak English,[19] and the physical mobility to lie on an examining table that did not accommodate mobility impairments,[20] as well as a degree of comfort in institutionalized spaces that were predominantly white and English speaking. All of these restrictions on accessing SACTCs helped to reinforce notions of the SAEK's ideal implicated user that were inscribed in the kit's forensic script.

The physical space in SACTCs reflected the priorities of the SAEK exam. Victims who were looking for other forms of counselling or support, such as group counselling or peer advocacy, had little hope of finding it in the institutionalized SACTCs. The examination room in the Women's College Hospital SACC featured a large examining table surrounded by other medical and forensic tools.[21] This physical space defined the kind of sexual assault treatment that the centre offered; sexual assault care and treatment in the SACC was a medical/forensic intervention that largely pivoted around the SAEK. The SACC's small consultation room could only comfortably accommodate a victim, a physician, and a nurse. The room size reflected notions of sexual assault treatment as an individualized exchange between medical personnel and victims. Sexual assault, within this physical arrangement, became an individual, medical problem, or as Doe (2003) says, "a personal tragedy rather than a social evil and a crime" (307). In 2000, the SACC opened an interview suite for police interviewing victims (Public Affairs, 2000). This structural addition solidified the connection between sexual assault treatment and the law that was embedded in the SAEK.

While SACTCs were creating and enrolling new experts into the SAEK's work, others were being pushed to the margins. When the Women's College Hospital SACC opened, questions emerged about the relationship between RCC advocacy, hospitalized sexual assault treatment, and the role of RCC advocates in SACTCs. A decade earlier, RCC advocates were often the only people in the emergency ward exam room who had received training on forensic exams and could offer guidance to physicians and nurses conducting the exam. However, with the introduction of the SACTCs and physicians and nurses who specialized in SAEK exams, the terrain was shifting. The SACTC's physical space allowed hospital staff to more stringently control RCC advocates' involvement in the forensic exam. SACTCs became a new site for "boundary work" (Gieryn, 1983, 781), in which hospital staff could draw and police lines between the new SAEK experts and RCC volunteers and staff. Hospital staff's boundary work situated RCC advocates as non-users of the SAEK who had limited influence over its use.

Within the first month of the SACC opening, the Toronto Rape Crisis Centre and the hospital were in negotiations over whether advocates had the right to accompany victims into the SAEK exam (Lipovenko, 1984). RCC advocates, according to the Toronto Rape Crisis Centre / Multicultural Women Against Rape (n.d.) collective, had long been "perceived as unprofessional, improperly trained, and threatening" (1), a sentiment that played out in the negotiations over advocates in the SACTCs. Two advocates I interviewed remembered these negotiations. One said, "We had to insist that we remain[ed] in the process" because, as another explained, "our role was not seen as a critical one." Hospital administrators expressed concerns that advocates were "not subject to the same controls as hospital staff" (Women's College Hospital, 1982, 2). The hospital acknowledged the value of volunteer rape crisis advocates, but proposed that "those volunteers selected to service this Centre be carefully screened, as it would be most damaging to individual victims as well as Women's College Hospital, if volunteers were to use the program to press personal beliefs regarding reporting and prosecution of rapists" (Women's College Hospital, 1983, 11). RCC advocates were positioned as risky outsiders who had the potential to negatively affect victims, the hospital, and the SACC. Despite this risk, the hospital did see the potential for developing a "working relationship" (12) with the RCC, if they agreed to promote the SACC. The hospital's SACC proposal read, "It is hoped that staff of the rape crisis centre would *recommend/convince* victims, who had not yet obtained care, to seek out the assistance of Women's College Hospital's Sexual Assault Treatment Centre" (12, emphasis added). Advocates' condition of entry into the SACC was to help advertise the SACC and the SAEK's forensic script to victims.

Several advocates that I interviewed remembered these tensions and described how difficult advocacy was in SACTCs, particularly for those volunteer advocates without professional credentials or training. One said:

The hospitals' inclination [was] to be more comfortable with people who were highly professionalized and some of our volunteers were not highly professionalized individuals who could show a card and talk the appropriate professional language. We were women helping other women ... The tension was based on the rape crisis centre's political analysis of rape ... intersecting with the hospital hierarchy and insistence on credentials ... Some nurses and doctors felt that rape crisis centre volunteers didn't have a role to play. In fact I think many of us were treated as though we were just in the way.

With the development of SACTCs and the professionalized expertise on the SAEK and sexual assault treatment that the centres housed, volunteer advocates could be more easily dismissed as non-experts who had little claim to being able to provide sexual assault care. The hospitals' "boundary work" (Gieryn, 1983, 781) and assertion of expertise aligned neatly with the provincial government's broader push to professionalize sexual assault advocacy and services.

Some RCCs questioned the rise of professionalized knowledge and practice around sexual assault and saw the SACTCs as an effort to undermine and co-opt sexual assault services and advocacy in the anti-rape movement. A third generation anti-rape activist recalled her grandmother's description of the SACTC opening in her Northern Ontario community. "I remember what my grandmother said when the SACTC came in ... 'This is the death now for women's groups. This is *the ultimate co-optation*'" (emphasis added). In line with this projection, the growth of SACTCs in Ontario in the late 1980s and 1990s was accompanied by an expanded mandate for them, which significantly overlapped with RCCs' work by including advocacy, public education, and research on sexual assault (Annual report, 1988–9). Some SACTCs adopted self-descriptions that mirrored the language that RCCs had used to define themselves years earlier (e.g., Why invest, 1979; Winner, 1977). One in particular read, "Our role is to help women regain control, power and choice in their lives. We play an advocacy role by influencing policy and requesting funding ... We are committed to practical and theoretical research designed to ... facilitate the understanding and prevention of sexual assault" (Interviews with women's health experts, 1999, 1).

Some RCCs spoke out against these developments by arguing that institutions, hospitals in particular, did not have the capacity to provide feminist advocacy for sexual assault victims, would transform sexual assault into a gender-neutral and medicalized issue, and would rob RCCs of their funding base. One centre wrote to the provincial NDP, which then formed the government, to urge them to continue funding RCCs despite the growing numbers of SACTCs, and to point to the dangers of hospital-based advocacy:

These services, while extremely valuable to their communities, are not able, nor do they have the expertise, to provide a comprehensive package of services informed by a feminist analysis of violence ... Generic services do not function to keep the issue of sexual violence in the public eye. Generic, gender-neutral services will medicalize, psychiatrize, and fragment services for those who have experienced sexual violence, and will serve to depoliticize this issue, eventually forcing it underground once more ... Traditional institutions cannot, or in some cases, will not, advocate for the social changes needed to get to the root of the causes of violence against women and children. Hopes for lasting social change come from a feminist analysis, and Sexual Assault Centres are, in many communities, the only feminist, gender-specific service. The NDP has favoured community-based services and has supported moving many services out of large institutions. Coalition centres have struggled over 20 years to make sexual assault a community issue – we are the voices who have spoken out and created change; we urge you not to participate in silencing us. (London Sexual Assault Centre, 1991, LSAC to ministers, 16 January 1991)

Despite these contestations, the SACTCs had solidified a new place for the SAEK and sexual assault treatment in hospitals, and in so doing had opened the doors to medical actors claiming a new form of expertise on sexual assault. The stage was set for the creation of a new professional group who would challenge the SAEK's forensic script by laying claim to the kit and sexual assault treatment.

Stabilizing with New Expert Users

As the numbers of SACTCs grew across the province, centre coordinators increasingly struggled to find willing physicians to staff them. They looked for alternatives to the physician-nurse model for the forensic exam and saw nurses as potential sources of cheaper and more readily accessible labour (Du Mont & Parnis, 2003). As a result, a new expert user for the SAEK emerged: the Sexual Assault Nurse Examiner (SANE). This new specialized professional redefined the kit's forensic script and contributed to silencing controversy, coordinating action, and stabilizing practices within the kit's network.

Sexual assault nursing first appeared in the United States, almost twenty years before the specialization developed in Canada (Ledray & Simmelink, 1997). In the early 1970s, American medical clinicians advocated for a new occupational group of nurses who would be responsible for the forensic exam (Donadio & White, 1974; Williams & Williams, 1973). These clinicians argued that nurses were uniquely poised to become the primary professionals for sexual assault treatment because of their extensive experience supporting victims during physician-led forensic exams and because the vast majority of nurses were women and could therefore more easily provide the necessary emotional care. The understanding of sexual assault care as inherently gendered work was thus at the foundation of the development of the sexual assault nursing profession. The first American Sexual Assault Nurse Examiner program was implemented in 1976 (Ledray & Simmelink, 1997). SANE programs in the United States multiplied in the years that followed, with their numbers dramatically increasing with the 1994 Violence Against Women Act (Fitzpatrick, Ta, Lenchus, Arheart, Rosen, & Birnbach, 2012). When SANE programs began to appear in Canada, they relied heavily on the model that had been established in the United States.

In 1992, nurses at a Ministry of Health conference expressed an interest in establishing a SANE program in Ontario. By that time, the Regulated Health Professions Act had expanded Ontario nurses' scope of practice to include the authority to insert instruments and hands "beyond the opening of body orifices" (Macdonald, Wyman, & Addison, 1995, 1-A),[22] which gave nurses the authority to take all the required samples in the SAEK exam. In the few years following, several Ontario nurses obtained SANE training in the United States, and in 1995 a SANE program was introduced to the SACC at Women's College Hospital in Toronto. The program allowed nurses to conduct SAEK exams without physicians if the victim did not meet any of the criteria for physician referral, which included conditions such as vaginal bleeding, pregnancy, and histories of psychosis and suicide attempts (Macdonald, Wyman, & Addison, 1995). According to Du Mont & Parnis (2003), the objective of the SANE program was to "increase efficiency, consistency, and quality of health care and evidence collection by using a single well-trained professional" (173).

Ontario SANEs built their professional status and claim to expertise by developing and asserting specialized knowledge on the SAEK, and through that, sexual assault and medical forensic science. In 1995, a SANE training program was introduced to Ontario, which included training on the kit, forensic evidence collection, and the techniques and skills for "maintaining professionalism and objectivity in medical documentation and courtroom presentation" (Du Mont & Parnis, 2003, 175). Through these training programs, SANEs solidified and learnt how to perform an expertise that other emergency room nurses and physicians did not have. They supported their claims to expertise with a wealth of empirical research on the benefits of SANE programs for improving forensic evidence collection (Ledray & Simmelink, 1997; Sievers, Murphy, & Miller, 2003), efficiency and quality of sexual assault treatment (Stermac & Stirpe, 2002), quality of sexual assault care (Dandino-Abbott, 1999), and prosecution rates in sexual assault cases (Aiken & Speck, 1995; Cornell, 1998; Hutson, 2002; Little, 2001). In 2007, Canadian SANEs formalized their professional status with the Forensic Nurses Society of Canada (Forensic Nurses' Society of Canada, 2012). Through these moves to professionalize and credentialize SANEs, the SAEK became the domain of a new professionalized expert.

Some physicians resisted the development of SANE programs in their hospitals. One sexual assault nurse described the forensic exam as "a physicians' domain initially." She recalled that when SANEs started to professionalize, physicians "did not want to let go" of the expert status that the SAEK exam had afforded them. Maier (2012) found that in the United States, some physicians "did not understand SANEs' role and resisted SANEs' presence in 'their territory'" (1326). Other physicians, however, reportedly worked cooperatively with SANEs and welcomed the reprieve of reduced responsibility for the forensic exam (Martin, DiNitto, Maxwell, & Norton, 1985). Reflecting these complicated historical tensions between physicians and nurses, many SACTCs in Ontario now rely almost entirely on SANE labour for forensic exams, but name physicians as their medical directors. These physicians provide SANEs in the program with general medical expertise when nurses require it and with the necessary medical directives that allow nurses to offer prescriptions to sexual assault victims that they could not otherwise. The medical hierarchy between physicians and nurses is thus preserved and reflected in the organizational structure of many SACTCs.

Despite their predominant role in SACTCs, SANEs' status as experts has sometimes come under fire in recent years. By taking responsibility for the kit exam, SANEs are called to testify in some sexual assault cases, sometimes as lay witnesses, where they report on the details of the SAEK exam, and sometimes as expert witnesses, where they express opinions on whether a victim's injuries could have resulted from consensual sexual activity. In Canadian law, expert evidence is only considered admissible if given by a "properly qualified expert" (R. v. Mohan, 1994).[23] In some sexual assault cases, defence lawyers have successfully argued that SANEs are not qualified experts as they have relatively limited medical training compared to physicians (R. v. Radcliffe, 2009), and unlike physicians, they lack experience examining non-sexually assaulted anatomy (R. v. Thomas, 2006). One defence lawyer I interviewed outlined his approach for challenging SANEs' expertise:

> I say things like, you've got no particular training in that area, you haven't been to medical school, you are not a physician, you are not an expert, you haven't conducted any studies, you are doing a sexual assault kit on people who are all coming in essentially indicating that there has been forced sexual activity in some way, so what kind of comparator is that?

In the courtroom, SANEs can thus be stripped of the expert status that they hold in SACTCs, while physicians' expertise is upheld. Although some SANEs have been deemed to be qualified experts in court and have given expert testimony, the fact that others' claim to expertise has been challenged points to how expertise is not a possession or a stable identity. Instead, expertise is always "situational" (Halfon, 2010, 70) and dependent on context. In the stabilizing medicolegal network, SANE's expertise was tied most clearly to the hospital exam room, where it was more often recognized and respected.

As SANEs completed more SAEK exams across the province, testified in more sexual assault cases, and SACTCs came to be known as the institutional homes for SANEs, the SAEK's forensic script shifted to accommodate the new actor.[24] Much of the clinical research on SANEs has emphasized their unique capabilities over emergency room physicians and other medical staff. This literature has argued that SANEs are more likely to (1) adhere to SAEK protocols than emergency physicians, (2) collect the stipulated amounts and different types of evidence the SAEK requires, and (3) store and label the evidence according to the SAEK's requirements (Campbell, Patterson, Bybee, & Dworkin, 2009; Campbell, Patterson, & Lichty, 2005; Dandino-Abbott, 1999; Ledray & Simmelink, 1997; Sievers, Murphy, & Miller, 2003). SANEs' adherence to the SAEK's script has been linked in the literature to increased rates of sexual assault prosecution (Campbell, Patterson, & Bybee, 2012; Campbell, Patterson, Bybee, & Dworkin, 2009; Ledray & Simmelink, 1997). Their work of ensuring that the SAEK's script is closely followed helps to situate the tool as a valued technology in medicolegal responses to sexual assault.

Through controversies and uncertainties about the new SAEK, SANEs and the SACTCs seemingly gave the SAEK's medicolegal network stability. The controversies that had surrounded the kit were seemingly silenced. New practices were developed to support its new role in medicolegal practice, and new actors who could be trusted to follow the SAEK's forensic script were put in place. Any lingering critiques from the 1970s that medical and legal institutions ignored rape were seemingly quelled by the new interest in and expertise on rape and sexual assault that actors within these institutions now claimed. The SAEK appeared to have been positioned as a trusted and largely accepted technoscientific actor that could produce reliable and credible testimony on rape. However, new scientific controversies, technological developments, and legal uncertainties were on the horizon. Forensic DNA typing was emerging in the late 1980s. With the rise of DNA typing, the SAEK's network would be destabilized once again and the kit itself would be reassembled alongside the new technology for identifying perpetrators of crime.

4 Assembling the Genetic Technoscientific Witness: Visions of Justice, Safety, and the Stranger Rapist

Look over your shoulder, we're coming. That's what DNA tells you. It's God's fingerprint.

Detective Lafreniere (as cited in Slade, 2011, B1)

In everyday litigation, as the material constituents of evidence are converted into scientific facts, their humble origins in the work of individual eyes and hands gets lost from view, and with this loss comes a forgetfulness about the shared social and scientific foundations of credibility.

S. Jasanoff (1998, 730)

When forensic DNA typing was introduced in Canada in 1989, many scientists, investigators, and government agencies touted the new technology as one that would revolutionize criminal investigations. For the first time in forensic history, proponents claimed, investigators would be able to identify unknown perpetrators of sexual assault and other violent offences from the traces they left behind at a crime scene. The semen, saliva, and blood on a sexual assault victim's body could lead investigators directly to the perpetrator of the crime. While the enthusiasm for the new forensic technique was high, the early 1990s were rife with debates in courtrooms, science journals, and government boardrooms about the reliability and credibility of DNA evidence. This chapter examines the turbulent rise of forensic DNA typing in Canada and reveals how it destabilized and restabilized the SAEK's medicolegal network. It reveals how, through these shifts, the Sexual Assault Evidence Kit was *re*assembled alongside DNA typing into a more credible tool for identifying perpetrators of sexual assault.

There are parallel stories of DNA typing's stormy history in other countries. Lynch, Cole, McNally, and Jordan (2008) examine the complex relationship between law and science that formed in the history of controversies around DNA evidence in the United States and the United Kingdom. Through their ethnographic and deeply theoretical narrative, they reveal how DNA profiling rose to be a "gold standard" (xiii) in forensic science that outmoded other forms of forensic evidence and gained the legal and science communities' trust to give truthful accounts of crime. Gerlach (2004) charts a similar history in Canada and explores the rapid acceptance of DNA typing in Canadian courts. DNA typing, for many Science and Technology Studies scholars, has served as a window into broader themes of credibility, truth, and sociotechnical and legal controversy. Work in this tradition has explored the seemingly unwavering credibility of contemporary DNA evidence in courts (Halfon, 1998; Jasanoff, 1998, 2006), the increasing commodification of DNA testing in private laboratories (Daemmrich, 1998), and the growth of DNA databanks (Cole & Lynch, 2006). I build on this existing literature by charting the tumultuous historical relations between DNA evidence, sexual assault, and forensics in Canadian sexual assault investigations. Through this focus, I draw out the connections between medicine, science, and law embedded in DNA typing in cases of sexual assault, and reveal the controversies over credibility and expertise within this history.

DNA typing radically changed the laboratory techniques for analysing the contents of the SAEK. In doing so, DNA typing redefined the kit, the evidence the kit produced, and the legal arguments that could be made based on its contents. As the kit was being reassembled, so was DNA typing. Forensic scientists, government agencies, and the media depicted the new technology as an ideal tool for sexual assault investigations.[1] They used cases of sexualized violence against women to frame DNA typing as a technology that would enhance public safety and protect the public, particularly women, from violent crime (e.g., Campbell, 1996; Curran, 1997; Gill, Jeffreys, & Werrett, 1985; Strauss, 1987). This framing helped propel the rapid acceptance of forensic DNA typing. The histories of DNA and the SAEK are thus intertwined. As DNA evidence was gaining credibility as a genetic witness of crime in the 1990s, the kit was gaining a new status as a genetic technoscientific witness of victims' bodies. By tracing the history of DNA typing, it becomes possible to see how the SAEK became accepted as a genetic technoscientific witness that could reliably facilitate the identification of sexual assault perpetrators.

The history of DNA typing importantly draws attention to how material things – blood, semen, DNA profiles, laboratory tools, and so on – formed new practices and relations between actors in the SAEK's medicolegal network. While these material things were present in the SAEK's network prior to the development of DNA profiling and blood and semen had "acted" in laboratories and courtrooms as evidence of rape in the 1970s and 1980s, it was not until DNA profiling was developed in the 1990s that blood, semen, and DNA profiles influenced the stability of the SAEK's network. These non-human actors became involved in destabilizing and rebuilding the SAEK's network as controversies flared about DNA typing and its use in sexual assault cases. This history of DNA typing thus highlights the role that non-human actors played in reassembling and stabilizing the SAEK and its network.

DNA typing moved much of the controversies around the SAEK from hospitals, where they had been in the early 1980s, into scientific laboratories, courtrooms, and government boardrooms.[2] It brought scientists, lawyers, and government actors into the centre of action around the SAEK and, correspondingly, pushed many of the rape crisis advocates even further to the margins. While some activists were involved in reforming sexual assault law in the 1990s, others were voicing criticisms of the increasing reliance on DNA technology. However, DNA typing proponents rarely heard their criticisms. DNA typing shifted the discussions around the SAEK and sexual assault from protocols, services, and treatment to bodily substances and identity. Accordingly, DNA typing swept particular actors into the centre of action around the SAEK and others onto the sidelines.

Forensic DNA typing emerged in a context of sexual-assault law reforms and new courtroom strategies for casting doubt on victims' credibility as reliable witnesses. This context set the stage for the introduction of DNA evidence as a credible genetic witness of crime and helped to enable rapid acceptance of DNA typing as a reliable technique for analysing SAEKs.

Victims' Credibility on Trial

The 1990s was a decade of significant change for both sexual assault law and the anti-rape movement in Canada. A right-wing conservative backlash had grown in response to the gains achieved by the women's

movement of the 1960s and 1970s. Many of the movement's achieve-
ments were being challenged or truncated altogether by conservative
social policies (Davis, 1999). Neoliberal economic policies of the 1980s,
which significantly reduced public funding for social services, continued well into the 1990s, and women's organizations across the country faced severe federal and provincial funding cuts (Pierson, 1993).
Rape crisis centres were no exception. In the 1990s, funding pressures
on these centres grew more acute, collective organizational structures
diminished, and the demands to professionalize rape crisis services
intensified. Under the weight of these pressures, advocacy work in
rape crisis centres was changing. Laws and legal practices for sexual
assault cases were also undergoing change. Many feminist lawyers,
scholars, and activists were organizing to implement legal reforms for
sexual assault, while defence lawyers were devising and solidifying
new defence strategies for challenging victims' credibility in court.

Sexual assault trials in the 1990s, according to some feminist legal
scholars, were "a minefield for female sexual assault victims" (Lee,
2000, 8), their credibility put on trial with new means and a new fer-
vour. The 1982 sexual assault law reforms, which had been intended
to improve victims' experiences in the criminal justice system, did little to mitigate this courtroom climate. In fact, the reforms prompted
a renewed commitment to challenge victims' credibility in court.
Defence lawyers sharply criticized the 1982 reforms for unfairly dis-
advantaging the accused and, in the words of one lawyer, providing
"a blue print for convicting people of sexual assault" (Schmitz, 1988,
43). They argued that in this climate new aggressive defence strategies were needed to challenge sexual assault victims' trustworthiness and credibility. Defence lawyer Michael Edelson proposed one
such strategy at a criminal bar meeting in 1988, where he asserted
that the defence should "whack the complainant hard at the preliminary ... If you destroy the complainant in a prosecution, you cut off
the head of the Crown's case, and the case is dead" (44).[3] The defence
could "destroy" a complainant, he claimed, by collecting discrediting evidence in a victim's psychiatric, hospital, and criminal records,
establishing a victim's drug use, and hiring a private investigator
to "beat the bushes and interview some of the principal Crown witnesses" (45). Rape crisis advocates, whom Michele Landsberg (2011)
interviewed in 1992, reported that Edelson's tactics were widely
followed in the 1990s. This was particularly so after two additional
legal reforms were implemented in the 1990s that were intended to

minimize the defence's interrogation into victims' sexual histories and personal records.

The 1982 sexual assault reform had instituted new restrictions on the admissibility of a victim's sexual history as evidence in the court-room. However, in 1991, the Supreme Court ruled that these restric-tions violated the constitutional rights of an accused under the Charter of Rights and Freedoms (R. v. Seaboyer; R. v. Gayme, 1991). Accord-ing to Johnson and Dawson (2011), many feminist advocates, scholars, and lawyers contested the ruling and argued that it would discourage women from reporting sexual assault. In response to this pressure, the Minister of Justice, Kim Campbell, organized a consultation of sixty women's groups, which helped draft new legislation on the use of sex-ual history (Stuart & Delisle, 2004). Landsberg recalled,

> It was an unprecedented, almost incomprehensible event: Conservative prime minister Brian Mulroney was in power, and here was his Minister of Justice, Kim Campbell, summoning grassroots and front-line women's groups ... I was among the startled invitees. Never before had a justice minister asked the advice of women in reforming the rape laws. (128)[4]

Out of these consultations, section 276 was added to the Criminal Code in 1992, which defined stricter rules around the admissibility of a vic-tim's sexual history. It was dubbed the new rape shield law. While this reform clarified the laws around the admissibility of sexual history, it did not preclude references to sexual history and authorized judges to rule on the admissibility of sexual history. A Department of Justice study in 1997 found that judges continued to admit sexual history as evidence in the majority of cases, often on the basis of vague defence arguments about relevance (Meredith, Mohr, & Carins Way, 1997).

In response to this new legislation, some defence lawyers turned to victims' personal records as a means of challenging their credibility in court and discrediting their testimony (Johnson & Dawson, 2011). In 1995, the Supreme Court ruled that personal records could be used if defence counsel proved that the records were relevant to the case (R. v. O'Connor, 1995). In 1997, Bill C-46 was passed, adding sec-tion 278 to the Criminal Code, which established a firmer set of require-ments for judges determining the relevance of personal records and included recognition of the victim's right to privacy. The reform was challenged shortly after as a Charter violation; however, the Supreme Court ruled in *Mills* (R. v. Mills, 1999) that it did not infringe on the

Charter rights of the accused and upheld the reform. Although some activists initially considered the Supreme Court ruling to be a victory, it did not stop many defence lawyers from continuing to argue that personal records were in fact relevant (Johnson & Dawson, 2011).

Despite some of the advancements in legal protections for victims gained in the 1990s,[5] many feminist scholars and activists argued that victims' credibility in the courts was continually undermined and challenged. In 1999, Penni Mitchell reflected back on the decade of legal reform and wrote, "No means no, but the war against sexism in the courts is far from over" (Mitchell, 1999). Against this backdrop, forensic DNA typing emerged as a new technology that promised to provide trustworthy evidence of sexual assault. As victims' credibility was being challenged in the courts, the credibility of the new genetic witness of sexual assault was being assembled. This assembly involved not only developing new laboratory techniques and technologies for analysing the SAEK, but also displacing old ones.

Making the SAEK's Contents Speak

Before the contents of the SAEK – the semen, blood, and hair samples – could become useful evidence in court, they had to be "made to speak" (Jasanoff, 2006, 330) with forensic laboratory techniques, tools, and scientists. Jasanoff describes how biological samples remain silent until they are made audible with scientific practices, technological interventions, and expert interpretations. It is through these processes, she argues, that biological samples gain authority as credible evidence in the courtroom. Applying this notion to sexual assault investigations, it could be said that the SAEK's biological samples remained largely silent until forensic scientists and technicians transformed them in the laboratory into evidence that police investigators, lawyers, judges, and juries could understand. It was a process of translation in which silent samples were transformed into evidence that then had the agency to act in the courtroom and describe parts of a sexual assault, its victim, and its perpetrator.[6] In the late 1970s and 1980s, kit samples were *made to speak* with forensic techniques that were geared towards excluding possible suspects of sexual assault. A suspect could be excluded if their physical characteristics did not match what scientists saw in samples from the SAEK.

Hairs in the SAEK were analysed most frequently with visual comparison methods.[7] Forensic scientists searched for visible matches between hairs in the SAEK and hairs collected from a suspect (Scientific evidence, 1980). A match could be used in court to suggest a suspect's guilt, but, according to one forensic scientist, matches were rarely found. Hair similarity, he recollected, was difficult to determine with any degree of certainty and could not be used to definitely identify a perpetrator. Hairs resisted being categorized and analysed in this way. Forensic scientists turned to techniques to analyse the other samples in the SAEK.

Blood and bodily fluids in the SAEK were analysed with blood typing. The ABO classification system was one of the most common systems used and involved identifying one of four blood types (A, B, AB, and O) in a sample (Eckert, 1978). Forensic scientists would associate each blood type with estimated frequencies of their occurrence in North American populations: type A was said to characterize 41% of the population, type B, 10%, type AB, 4%, and type O, 45% (Scientific evidence, 1980). To specify the analysis, some scientists used protein and blood grouping systems, in which particular combinations of blood types were used to formulate smaller population frequencies (Krishnan, 1978). Forensic scientists used the blood type or grouping that they identified in a SAEK sample to exclude all suspects with that blood type (ibid.). Like hair, however, blood could not definitively reveal the suspects' identity. These techniques, according to some forensic scientists, "left a large degree of uncertainty" (Gill, Jeffreys, & Werrett, 1985, 578) that often played out in the courtroom.

An Ontario Crown prosecutor remembered that in the 1980s, forensic hair and blood analysis was rarely useful in sexual assault cases. Recalling this with frustration, he said, "It was almost impossible to get a conviction ... because you didn't have the scientific evidence linking the accused to the victim." The available forensic techniques could only reveal a *possible* link between an accused and a victim if a match was found between blood or hair types. He described blood grouping by saying, "It was very very very weak evidence ... That evidence was so weak because of the vast number of people who are in blood groups. The evidence was *negligible* [emphasis added]." Despite the promises in the early 1980s that the new SAEK would be Ontario's greatest "rapist trap" (Crawford, 1984, A13) and would secure successful convictions of rapists across the province, it had in fact proved to have little evidentiary value in the courtroom. While the kit could reveal a victim's injuries after a sexual assault, it offered little insight into who the perpetrator was. The stage was thus set for a new actor to enter the SAEK's medicolegal network.[8]

In 1985, Alec Jeffreys, a geneticist at the University of Leicester, England, and his colleagues announced that a new forensic technique would replace earlier forensic methods for identification and "revolutionize forensic biology" (Gill, Jeffreys, & Werrett, 1985, 577; Jeffreys, Wilson, & Thein, 1985). They asserted that this new technique, which they dubbed DNA fingerprinting,[9] would allow scientists to identify individuals with more precision and certainty than any previous forensic biological procedure had offered. They claimed that DNA fingerprinting produced a genetic code from biological samples that was specific to an individual, much like the patterns of actual fingerprints were claimed to be (Cole, 2001). Perpetrators of crime, they asserted, could be identified from the biological fluids they left at the crime scene.

To produce a DNA fingerprint, Jeffreys and his colleagues used a multistaged process that resulted in an image of variations in DNA sequences that were said to be specific to individuals. Restriction Fragment Length Polymorphisms (RFLP) analysis began with the chemical extraction of DNA strands from a bodily fluid sample. The DNA strands were then cut into fragments with enzymes, an electric current was applied to arrange the fragments by size, radioactive probes were applied to targeted DNA, and the fragments were blotted onto a membrane from which an image of the DNA fragments was produced. A bar code-like image of DNA fragments resulted, which was considered the DNA fingerprint. If a suspect's DNA fingerprint matched the DNA fingerprint developed from a sample at the crime scene, it could be argued in court that the suspect was likely responsible for the crime. This new technology seemingly held the promise of radically reshaping the laboratory techniques for analysing all crime scene evidence, including SAEKs.

Enrolling a New Actor in Rape Investigations

From the beginning, DNA profiles were cast as powerful new actors for rape investigations. In 1985, Gill, Jeffreys, and Werrett forecasted that DNA typing would revolutionize the analysis of forensic samples for all violent crime, but particularly for rape cases. They envisioned that it would overcome the limitations that had plagued hair and blood analysis techniques in rape investigations in the past, and allow scientists to extract sperm nuclei from vaginal fluid. For the first time in forensic biology, they claimed, scientists would be able to give a "positive identification of the male donor/suspect" (Gill, Jeffreys, & Werrett, 1985, 577) from semen samples.[10]

Before forensic DNA typing could become the tool that Jeffreys and his colleagues envisioned for sexual assault investigations, it had to become part of the network of actors involved in rape case investigations and be accepted as a trusted tool for identifying perpetrators of violent crime. Building trust in this new actor would also mean displacing others. Hair and blood analysis techniques that had been used in the past would have to be rendered less credible and less valuable for analysing rape case evidence. Branding forensic DNA typing as a trusted technique for making the SAEK's contents speak would require the work of not just scientists, but also lawyers, police investigators, government agencies, and mass media. Jeffrey and his colleagues' announcement of forensic DNA typing marked the beginning of what became a decade of struggle to build, challenge, and cement the power of forensic DNA analysis in rape and sexual assault investigations.

Not long after Jeffreys and his colleagues penned their predictions for DNA typing, the new technique became a key player in a sexual assault murder investigation in Leicester, England. In 1986, two women named Lynda Mann and Dawn Ashworth were sexually assaulted and murdered by an unknown perpetrator (Jail term cut, 2009). Investigators conducted DNA testing on 5500 blood samples collected from men in surrounding communities (Wambaugh, 1989). After some initial challenges with sampling, a match was eventually found, and Colin Pitchfork was confirmed as the perpetrator of the crimes and convicted. He was the first person to be convicted of a crime based on forensic DNA evidence.

The news of Gill, Jeffreys, and Werrett's work and its application in the Leicester case spread quickly to North America. Many Canadian and American forensic scientists were vocal about DNA's potential to transform sexual assault investigations and SAEK analysis. In 1987, John Wilker, the president of a private American forensic lab, asserted DNA's potential by saying, "In rape cases in which sperm can be recovered, there is just no way that could be analyzed until now" (Hilts, 1987, C16). In Canada, Douglas Lucas, the director of the Ontario Centre of Forensic Sciences (CFS), echoed this by saying that DNA fingerprinting's "greatest application would be in sexual assault cases" (Strauss, 1987, D4).

The excitement about DNA typing in sexual assault investigations was situated in a rush of media coverage in the late 1980s about DNA's more general potential to transform forensic investigations of all violent crimes. Canadian and American media were quick to characterize DNA typing as a radically new, infallible technique for identifying perpetrators of crime. Canadian newspaper headlines described DNA as the "hot new crime buster" (MacCharles, 1988, A1) that "points at the guilty" (Strauss, 1987, D4) and "makes criminal identification certain" (Hilts, 1987, C16). DNA analysis was proclaimed to be the "biggest advance in the science of crime detection in a century" (Lohr, 1987, G8). In some reports, DNA analysis was likened to a divine intervention for crime detection. One detective in Calgary, Alberta, reportedly said, "It's like God opened up the clouds and said to offenders, 'Stand by, I've got a new process for the police to use'" (DNA prints big advance, 1987, H8).

By 1988, the Ontario Centre of Forensic Sciences was receiving dozens of DNA typing requests from police and prosecutors across the country (MacCharles, 1988, A1). To meet the increasing demand, the Royal Canadian Mounted Police (RCMP) forensic laboratory was given a $3 million dollar budget to begin the preparations for a DNA typing laboratory, and the Ontario Centre of Forensic Sciences was allocated $100,000 for DNA typing research (ibid.). The growing anticipation of the technology's arrival in Canada was, however, not universally shared. Several scientists and lawyers expressed hesitancy about how the courts would receive the new technology. The CFS director was reported as saying that despite the growing excitement about DNA typing, "forensic scientists were laying low ... Nobody wants to rush before a judge and jury until the method is known to be foolproof" (MacCharles, 1988, A1). Several lawyers similarly voiced their concerns and questioned the feasibility and ethics of obtaining DNA samples from unwilling suspects (Dunlop, 1987, C1).

Despite any hesitations there may have been, in the fall of 1989, the RCMP laboratory in Ottawa began accepting DNA samples (Counsell, 2007); in July 1990, the Centre of Forensic Sciences followed suit (Campbell, 1996). In the five years after, DNA typing was used in over 1000 criminal trials in Canada (R. v. S. F., 1997). While the sheer numbers of cases using DNA typing implied a general acceptance of the technology, the first few years of forensic DNA typing were turbulent. Legal and scientific controversies around DNA typing were rife in academic journals and courtrooms, while scientific practices and technologies for DNA typing were rapidly changing. The entrance of forensic DNA analysis into the medicolegal network of practices for rape and sexual assault investigations was not as smooth as some had predicted.

A Rocky Beginning: DNA and SAEK Evidence

Concerns about RFLP DNA typing in sexual assault cases spread to the forensic science community in the late 1980s and 1990s. Reynolds, Sensabaugh, and Blake (1991) argued that although RFLP DNA analysis had made a significant mark on forensic science, its application in sexual assault cases was marred with challenges. The significant amount of time that RFLP DNA analysis required caused delays in investigations, they asserted. In addition, the large amount of non-decayed blood or semen needed to conduct RFLP analysis was rarely available in sexual assault cases, where samples were far more likely to be small and decayed.[11] These critics were not alone. A few years earlier, some molecular biologists had argued that RFLP DNA typing could not easily be done on samples from sexual assault kits in the 1980s. In 1987, David Houseman, a molecular biologist at MIT, contended that kits from the 1980s were designed for older forensic techniques that required only small amounts of semen to confirm that sperm was present (Bass, 1987, A1). In contrast, he asserted, the large amounts of sperm that RFLP DNA testing required was rarely collected with the existing kits. While the stage was set for discussions on redesigning the SAEK, much bigger changes in practice were in store. Scientific practices around DNA typing were changing in the late 1980s with the development of Polymerase Chain Reaction (PCR) and Short Tandem Repeat (STR) analysis, two new techniques for DNA typing.

PCR is a technique that is now used to duplicate the amount of DNA in a sample with a series of heating and cooling cycles. With PCR, millions of copies of a DNA molecule can be produced within a few hours, a process said to be akin to "molecular Xeroxing" (Curran, 1997, 8). This can reduce the size of sample required for DNA typing from the width of a 25-cent coin to 0.3–0.5 nanograms (a billionth of a gram). When PCR was first introduced to Canada in the early 1990s, it was reported as a "huge leap for the science" (Kozicki, 2007, 42) that would expand the capacity of forensic DNA typing and significantly reduce the time it required.[12] The introduction of PCR was accompanied by the development of Short Tandem Repeat analysis, which was focused on analysing short repeating segments of DNA that vary in length between individuals. Using PCR amplification on STR loci allows scientists to generate DNA profiles from significantly smaller and potentially degraded samples collected with the SAEK.

These changes enhanced the usability of the SAEK, according to the forensic scientists I interviewed. They asserted that STR and PCR techniques significantly increased the likelihood that scientists could devise a DNA profile from the SAEK contents. While these developments seemingly gave DNA typing some renewed credibility in sexual assault investigations, heated controversies circulated in courtrooms and science journals in Canada and the United States about the value and credibility of DNA evidence in all criminal investigations.

The early to mid-1990s in the United States were characterized by what many have called the "DNA wars" in criminal trials (Thompson, 1993, 22). The American DNA wars centred on the reliability of DNA collection and analysis and the accuracy of DNA profile interpretation. While some prosecutors and forensic scientists avidly advocated admitting DNA evidence in criminal trials, others argued that the flood of excitement around DNA had led courts to prematurely rush to introduce DNA evidence before the scientific community had deemed it to be reliable (Hoeffel, 1990; Lander, 1989). Legal scholar Janet Hoeffel argued in 1990 that DNA had been "steamrolled ... through the courts" (466) and that restraint must be exercised "on the acceptance of unproven novel scientific techniques that turn courtrooms into laboratories and defendants into guinea pigs" (467). The debates that ensued were aired in US courtrooms[13] and in published articles in science and legal journals.

In Canada, the so-called DNA wars were far less animated. Canadian courts were less concerned with the technical reliability of DNA analysis than their American counterparts and focused instead on the interpretation of DNA profiles with probabilities of identification (Gerlach, 2004). All forensic DNA profiles were interpreted with statistical estimates of the likelihood that individuals other than the suspect could match the DNA profile in the SAEK or at the crime scene. The estimates were based on databases of DNA that were intended to represent the genetic variability within a population, but were at the time loosely compiled from blood banks, police volunteers, and other non-random samples and divided into roughly defined racial subgroups. Some population geneticists challenged the validity of statistical estimates based on the early databases and raised concern about "the subpopulation problem" (Cohen, 1990, 358), which highlighted the limitations of statistical estimates that were based on incomplete data on DNA profile frequencies in subpopulations.

In Canada, questions around the subpopulation problem first appeared in a sexual assault murder case in British Columbia in 1991 (R. v. Baptiste, 1991). The defence argued that the DNA evidence derived from the vaginal swab and a bloodstain on the accused's pants had been based on insufficient subpopulation data on the small Indigenous community in British Columbia where the suspect was from. They asserted that, as a result, there was "no scientific basis" to the scientist's conclusion that the possibility that the DNA came from someone other than the suspect was extremely remote. DNA's capacity to reveal identities of sexual assault perpetrators was being put on trial. Despite questions raised in court, the trial judge and the Court of Appeal both ruled that the DNA evidence was in fact admissible. The subpopulation problem continued to appear in some subsequent Canadian cases, until the late 1990s when the debates subsided with improvements to genetic data for populations (Frederiksen, 2011).[14]

By the mid-1990s, the most vocal critics and proponents of DNA typing were proclaiming that "the DNA fingerprinting wars are over" (Lander & Budowle, 1994, 735). As Lynch et al. (2008) trace in their historical narrative, American controversies over DNA evidence had been significantly quieted by numerous consultations on DNA evidence in the early 1990s, the subsequent standards and protocols that had been developed, and the release of the National Research Council report on forensic DNA evidence in 1996. In a symbolic move, two opposing participants in the DNA controversies, Eric Lander, a key critic of forensic DNA typing, and Bruce Budowle, a proponent of DNA typing, asserted together in 1994 that there was "no remaining problem that should prevent the full use of DNA evidence in any court" (Lander & Budowle, 1994, 735). They called for a widespread acceptance of DNA evidence: "The public needs to understand that the DNA fingerprinting controversy has been resolved. There is no scientific reason to doubt the accuracy of forensic DNA typing results ... Now, it is time to move on" (738).

While the DNA wars in American and Canadian courtrooms had subsided, questions about the use of DNA typing in sexual assault investigations had not. Laboratory practices around DNA evidence came under fire in Ontario in the early 1990s in the wake of the highly publicized Paul Bernardo investigation for serial sexual assaults and murders of young women. This case triggered a public inquiry into the failings of the provincial police and the Ontario forensic laboratory to identify and apprehend Bernardo. This case cast critical light on the efficiency of DNA testing and the potential of its mishandling in sexual assault investigations. More importantly, however, the case helped to situate DNA typing in the public imagination as a crucial technology for sexual assault investigations. It also reinvigorated former arguments against the older forms of evidence from blood grouping and hair analysis. By drawing attention to the failures of DNA analysis, this case inspired counter-claims that DNA analysis could reliably identify perpetrators of violent crime and protect women from sexual perpetrators.

In 1996, Justice Archie Campbell released a report on the institutional failings in the Paul Bernardo investigation (Campbell, 1996). He described how, between the years 1987 and 1992, police had failed to identify Bernardo as the perpetrator of eighteen sexual assaults and three murders of women in Ontario. His inquiry revealed that Bernardo remained unidentified for years because of delayed forensic analysis and a series of disjointed, uncoordinated, and often disinterested police investigations. In 1987 and 1988, police had submitted multiple SAEKs for blood grouping analysis from women whom Bernardo had raped. Campbell documented how, after a few failed attempts, the blood grouping results led police to reduce the suspect pool to 20% of the male population and, later, to 13% of the male population, neither of which was helpful in identifying Bernardo. Campbell stressed in his report that "conventional" (43) blood-grouping tests were inadequate for identifying assailants with any precision. He recounted how CFS had ignored the police request to analyse Bernardo's DNA when it was resubmitted with ninety-two other samples from possible suspects. It was not until 1992, two years after the first SAEKs were sent to CFS, when the police resubmitted the samples for DNA testing, that the analysis was actually completed and Bernardo was identified and later arrested. According to Campbell, during the time Bernardo's DNA was sitting at CFS unanalysed, Bernardo raped four women, and tortured and murdered two others.

Campbell concluded that if Bernardo's DNA samples had been analysed from the SAEKs when they were first submitted in 1990, the rapes and murders between 1990 and 1992 would not have occurred. He framed DNA analysis as an innovative technology that, if used in a timely way, could prevent violence against women, unlike the "conventional" (43) techniques of blood and hair analysis. Delays in

DNA analysis were cast as a risk to public safety and a justification for increased spending on DNA analysis. Campbell wrote, "The Bernardo case demonstrates that delays in DNA testing can imperil personal safety and cost lives. Any reluctance to continue to spend the public funds necessary to maintain a reasonable turnaround time for DNA tests must give way to a consideration of the financial and human cost of failing to do so" (9). Campbell's simple formula for ensuring public safety with DNA analysis of SAEK evidence was echoed in the media, where DNA typing was described as the tool that *could have* protected some of the women that Bernardo raped and murdered if it had been used earlier in the investigation (Bernardo DNA sample, 1996; Kirk, 1995). Even one of Bernardo's own lawyers was quoted as saying, "If they took samples … and [had] done the proper tests, I think none of this would have happened" (Kirk, 1995, A1).

Campbell's report revealed that long before Bernardo's DNA was submitted to CFS for analysis, several women came forward to police reporting Bernardo's likeness to the composite drawing of the Scarborough rapist and his aggressive and violent behaviour towards women. Investigators did not follow up on any of these tips. Campbell attributed these failures to a lack of adequate policing staff and of a computerized management system for organizing information on the case. More systemic issues of police failing to take women's reports of violence seriously were notably absent in Campbell's report. Despite the potential that these tips could have led police to Bernardo in the early stages of the investigation, it was DNA, not better police practice, that the media touted as that which could have prevented many of Bernardo's crimes.

While DNA was heralded by some as the potential saviour of women from violent crime, others were less convinced. Some feminist writers criticized Campbell's misplaced enthusiasm for technological solutions to violence against women and his failure to recognize the systemic sexism that characterized police responses to sexual assault investigations. Martin Dionne (1997) criticized Campbell's report by saying, "He would have us believe that technology is the policing cure" (3) for failed investigative techniques and violence against women. Technological solutions to violence, Dionne argued, "simply buttress and perpetuate a system of law enforcement that attaches little significance to the experiences of women who have been raped, that fails to listen to their evidence, and that relies on stereotypes of both women and the men who rape for its investigative paradigm." These critiques appeared to gain little traction in mainstream media. Campbell's report seemingly had done its work to situate DNA typing as a necessary technology for protecting women from violent crime.

Continued Contestations

Despite what seemed to be a growing acceptance of DNA evidence as an indispensable actor for sexual assault investigations, critics remained. In 1994, academic feminist Helen Holmes posed the question "Who benefits from the development and use of DNA typing?" (Holmes, 1994, 232). Through a detailed analysis, she observed that DNA typing had brought police and Crown attorneys newfound credibility, and forensic laboratories increased legitimacy, secured employment, and funding. Victims, however, were rarely benefiting from DNA typing, she argued. Kubanek & Miller (n.d.) penned a similar argument that DNA typing was being falsely promoted in women's interests, despite rarely proving itself useful for women, particularly in cases of sexual assault and rape. Most sexual assault cases, they noted, involve perpetrators whom a victim knows, such as a father, brother, uncle, friend, boyfriend, or husband. In these cases, the legal question is more often one of consent, not of identity. Therefore, they reasoned, a technology for identifying the perpetrator of the attack is superfluous more often than not. In the few cases involving stranger rapists, Kubanek and Miller predicted, DNA evidence would encourage more defendants to admit to sexual activity with the victim and argue that it was consensual and therefore not criminal. Consent defences, according to Kubanek and Miller, had been proven to be far more challenging for the prosecution. The assertion that DNA typing was a necessary technological intervention for protecting women and girls was merely "pander[ing] to the myth of the dangerous stranger" (2). DNA evidence, Kubanek and Miller also stressed, threatened to increase police and courts' expectations of scientific corroboration of women's reports of sexual assault and decrease the value of their testimonies. They wrote:

> Increasing the weight given to scientific evidence subtly alters the notion of reasonable doubt ... DNA evidence provides such a high level of statistical reliability that other types of evidence pale in comparison ... In cases where DNA evidence is not available ... the usual evidence accumulated against the accused may look weak. In the eyes of the judge and jurors, the verbal testimony of witnesses, especially that of the victim, cannot carry

the statistical reliability of scientific evidence, a bias which can only work against women in the majority of cases. (2)

For Holmes (1994) and Kubanek & Miller (n.d.), DNA evidence did not have the power to shield women from violence. Instead of deterring men from raping women, Holmes argued, DNA evidence was more likely to inspire new strategies among rapists to obscure or eliminate identifying forensic traces of rape, such as wearing condoms or participating in gang rapes. Instead of preventing violence, Kubanek and Miller forecasted, DNA evidence would have the unintended effect of making cases without DNA evidence seem "unconvictable" (2), which would actually decrease the number of convictions of sexual assault in Canada. DNA evidence, they argued, was more likely to give government agencies the false impression that they had solved the problem of sexual assault. Kubanek and Miller (n.d.) concluded by saying, "These interventions cannot be defended in the interests of women subjected to male violence ... We will not have these actions taken in our name" (4).

Despite some continued critique and resistance, by the early 2000s, DNA analysis had seemingly solidified a new status as a reliable and necessary forensic technique for analysing the SAEK. This status would be strengthened even further with new developments that increased the use of DNA typing for identifying sexual offenders.

Extending DNA's Reach

In the late 1990s, Canadian government agencies were making preparations to expand the use of DNA typing in criminal investigations. The resulting changes would significantly alter the way that DNA typing was used in all criminal investigations, including sexual assault cases. Before these changes were put in place, a DNA profile from a SAEK was only useful to investigators if they had a DNA profile from a known suspect to compare to the profile found in the SAEK.[15] Without a known suspect, the DNA profile from the SAEK could not be linked to any particular individual and remained an anonymous profile. All of this changed in 2000 with the establishment of the National DNA Databank. Similar to the DNA databanks that had been developed in the United Kingdom and the United States in the mid-1990s, the Canadian DNA databank allowed investigators to compare unknown DNA profiles from SAEKs to banked DNA profiles from individuals previously convicted of a crime and ordered to submit a DNA sample. If a match was found, investigators could for the first time assign a known identity to an unknown DNA profile from a SAEK.

The DNA Identification Act (1998) authorized the construction of the new databank in 1998. With a $10.9 million dollar start up budget, the National DNA Databank opened in June 2000 (National DNA Databank, 2003). The databank was situated in Ottawa, run by the RCMP, and housed two indices: a convicted offender index, which would store DNA profiles from convicted offenders court-ordered to provide a DNA sample, and a crime scene index, which would store unknown DNA profiles collected from crime scenes and SAEKs (National DNA Databank, 2006). Within two years of being in operation, the databank amassed 10,261 DNA profiles (National DNA Databank, 2003), a number that has since swelled to 470,223 as of February 2017 (Royal Canadian Mounted Police, 2017).[16]

Like DNA in the 1990s, the DNA databank was characterized as a necessary tool for enhancing public safety. The Solicitor General's first press release introducing the databank asserted that the bank would "give us a powerful investigative tool that will protect Canadians from violent criminals. It will help ensure that those guilty of serious crimes, such as repeat sex offenders and violent criminals, are identified and apprehended more quickly" (as cited in Curran, 1997, 24). RCMP commissioner Zaccardelli echoed these claims in the first annual report of the databank, where he emphasized the databank's value: "For Canadians, it means safe homes and safe communities" (National DNA Databank, 2002, 2).

Not all people in Canada welcomed the introduction of the DNA databank with such eagerness. Some criminal lawyers expressed concerns that the bank "put too much personal information in the hands of the state" (van Wageningen, 2000, A1). The Privacy Commissioner of Canada (1995) voiced concerns about the large number of criminal offences captured in the databank. Large numbers of women's groups also criticized the bank. According to Kubanek and Miller (n.d.), 670 member groups of the National Action Committee on the Status of Women opposed the development of a National DNA Databank. Voicing this opposition, Kubanek and Miller argued that banking offenders DNA was being "falsely romanticized as a 'quick fix' for the systemic problems that women face in dealing with the criminal justice system" (3). They questioned the databank's potential to reinforce inequalities in Canada and contended that the databank would inevitably reflect the disproportionately high conviction rates for men of colour, Indigenous men, and poor men, and would aid the continued criminalization of these groups. They attacked the databank's hefty budget and questioned how it was that women's advocacy organizations were so poorly funded when there were available funds for enhancing the strength of forensic DNA analysis.

Feminist critiques of the databank, according to Gerlach's (2004) historical analysis, were met with little interest from government agencies. In one speech, the Solicitor General briefly noted the women's groups concerns, but quickly dismissed them by reinforcing DNA as essential for public safety,

> There are those who believe that resources earmarked for a DNA databank would be better spent on family violence programs or women's shelters. In response, I would agree that spending in those areas is important ... but at the same time, there is no doubt in my mind that a national DNA databank will add to the safety of all Canadians (as cited in Gerlach, 2004, 88).

This is one of the few documented government responses to feminist critiques of the databank.

Through the debates about DNA typing and DNA data banking, proponents had maintained a simplistic equation between DNA typing and public safety. Media, government agencies, legislators, and scientists drew on women and girls' experiences of sexualized violence to bolster the strength of this equation. The National DNA Databank (2002, 2003, 2006) annual reports routinely employed stories of violence against women and girls to illustrate DNA's powers: nearly all the case examples of DNA identifying and convicting offenders in these reports were stories of male violence against women and girls. One such case involved a man who had sexually assaulted a young woman in 1998 and was identified several years later when the databank matched his DNA profile to the one found on the victim's body. The annual report read:

> He would likely be a free man today if it were not for the National DNA Databank ... According to the detective who led the police investigation, "It shows the money spent on the database is leading to police finding and convicting sexual predators, where we have very little evidence. Without the National DNA Databank, this predator would still be out there in our midst. (National DNA Databank, 2003, 32)

Embedded in these narratives of the databank's successes was a threat of the imminent danger of unknown sexual predators "in our midst" (32). The databank and DNA were configured as the only tools that could combat this danger of the stranger rapist and ensure that innocent women and girls were protected and, if victimized, given justice. These case examples propagated tired myths of sexual assault as a crime most commonly committed by strangers who could only be identified with the databank. Most importantly, they capitalized on public fears of gendered violence to advertise the powers of DNA and the databank. As a result of these and many other efforts to draw attention to DNA's powers, DNA typing had gained a prominence in the criminal justice system as a reliable, credible, and necessary forensic technique for identifying perpetrators of sexual assault.

Reassembling the SAEK

DNA typing and the National DNA Databank had radically altered how the SAEK's contents were *made to speak*, and reassembled relations in the SAEK's medicolegal network. These new forensic techniques transformed the laboratory practices that forensic scientists used to produce evidence from the SAEK, and redefined the type of evidence that scientists could produce from the contents of the SAEK. Outside of the laboratory, DNA typing had redefined the SAEK itself. The acceptance of DNA typing and the development of the databank had redefined the SAEK as a tool that could theoretically identify perpetrators of sexual assault. The new laws and institutional practices around DNA typing reassembled the SAEK's meaning and purpose, and the kit became a DNA evidence collector. In the early 2000s, the Ontario Centre of Forensic Sciences began efforts to redesign the SAEK's material form to suit the needs of its new purpose.

The Ministry of the Solicitor General organized a SAEK Working Group in the early 2000s, which was charged with the task of reviewing the SAEK and its impact on victims of sexual assault.[17] This was largely in response to the growing recognition among many medicolegal experts that the kit exam was, for many women, "long, humiliating, and almost as traumatic as the sexual assault itself" (Griffiths, 1999, 96), as Jeffrey Griffiths described in his audit on sexual assault investigations in Toronto. Alongside the SAEK Working Group, CFS conducted its own review of the kit in 2001 to reconsider the kit "in light of current

DNA technology."[18] Findings from both the SAEK Working Group and the CFS SAEK review were used to redesign a new SAEK.

The CFS released the new kit in Ontario sexual assault treatment centres in 2001. In educational literature for the new kit, they praised its benefits, claiming that it would be a "less intrusive procedure for [the] victim," would ensure "faster analysis within the lab [and] facilitate storage," and have an "optimal design for DNA analysis," which would create "more DNA-ready samples."[19] Although the redesigned SAEK was not significantly different in material design from its previous iteration, the kit's new alliance with DNA typing fuelled claims that it was in fact a radically improved technology.

The new kit added and eliminated a few evidence collection steps and included several new technologies for storing biological evidence. To allow forensic scientists to distinguish between a perpetrator's DNA and a victim's, the new SAEK included a buccal swab for obtaining a DNA sample from a victim's cheeks, tongue, and gums.[20] Several steps in the SAEK exam were also removed, in response to the decreasing use of hair and blood typing analyses.[21] Plucking pubic hairs, which had a decade earlier caused heated contention between some scientists, physicians, and anti-rape activists (Martin, DiNitto, Maxwell, & Norton, 1985; Meeting notes, 1983), was no longer included in the SAEK exam. One forensic scientist I interveiewed recalled this decision and said, "It was used so infrequently that the pain and the trauma [was] not worth it for those rare cases where you m[ight] find something." The vaginal aspirate was also removed from the exam for similar reasons, as swabs had proven just as effective for collecting seminal fluid in women's vaginal cavities. One sexual assault nurse cheerfully remembered when the vaginal aspirate was eliminated from the SAEK, and said that vaginal aspirates were "messy [and] didn't really give us a whole lot of evidence. Therefore they decided it wasn't worth it, with nurses, and I know, patients rejoicing [laughs]."

Most significantly, the new kit was designed to be stored at room temperature.[22] Previously, all SAEKs had been frozen. However, a forensic scientist I interviewed said that it was common for samples to spoil, leak, decay, or be damaged when the frozen kits were transported from hospitals to police evidence units to the CFS. A new container, the "RT swab box,"[23] developed to allow moist swabs to breathe and dry, was included in the new kit. The change to a self-drying SAEK made it possible for hospitals to store kits while victims decided if they wanted to report to the police. As a result of this technological

Figure 4.1 Training document for revised SAEK, 2001 (© Queen's Printer for Ontario, 2001. Reproduced with permission. The materials are current to 2016 and are subject to change.)

CENTRE OF FORENSIC SCIENCES · CFS · Est. 1951 · CENTRE DES SCIENCES JUDICIAIRES

IMPACT OF KIT IMPROVEMENTS

	Complainant	Nurse Examiner	Police	CFS	Legal
LESS INTRUSIVE PROCEDURE No Aspirate, FTA buccal swab	✓	✓			
IMPROVED TIMELINES Less Unnecessary Sampling	✓	✓		✓	
STEPWISE COLLECTION MODULES	✓	✓			
DRY STORAGE - Biology REFRIGERATION - Tox.		✓	✓	✓	
PRESERVATIVE FOR URINE	✓			✓	✓
CHECKBOX CFS SUBMISSION FORM			✓	✓	
KIT OPTIMIZED FOR FASTER ANALYSIS - Smears	✓		✓	✓	✓
KIT OPTIMIZED FOR DNA	✓		✓	✓	✓

development, a new policy was introduced in the hospital treatment centres that allowed them to store SAEKs for up to six months while victims decided if they wanted to report to police. SANEs noted this as one of the most significant advancements in hospital policy for victims of sexual assault.

The reassembled SAEK was seen as having significantly more value than its predecessor. One police officer remarked, "The introduction of DNA has been the biggest advancement in terms of catching a perpetrator ... The value of the kit has changed because of DNA ... [It] becomes more valuable because it contains DNA." The reassembled SAEK and the network of practices and actors around DNA typing made the DNA in the kit samples perceptible and knowable, a change that dramatically increased the perceived value of the kit itself.

The Reassembled SAEK in Practice

The vast majority of sexual assault police investigators that I interviewed equated DNA testing with the SAEK, as if in the past decade, the two had become synonymous. Before DNA, the kit was understood as a tool that could, at best, *eliminate* suspects, whereas now the kit had become a tool for *identifying* suspects, in addition to collecting corroborative evidence of physical injury. DNA typing shifted the primary emphasis of the SAEK exam to gathering legal evidence of identity, despite the available crime statistics at the time that suggested that 80% of sexual assaults reported to police in Canada involved perpetrators whom a victim knew (Kong, Johnson, Beattie, & Cardillo, 2002). When the SAEK was put to use in medicolegal practice, it conveyed a new emphasis on identity that reflected and reinforced a dated myth of sexual assault as an act perpetrated by strangers.

Forensic DNA evidence has been dubbed in the public imagination as *the genetic witness* of crime (Aronson, 2007). This moniker conjures up images of invisible genetic material acting as a "silent but credible witness" (National DNA Databank, 2003, 26) of criminal activity. Allied with this new actor, the SAEK in medicolegal practice witnesses the victim's body – or, as it is often conceived, the sexual assault crime scene – in new ways. The reassembled kit searches the crime scene for traces of DNA. The SAEK, with DNA typing, has thus become a new type of technoscientific witness of sexual assault: a *genetic* technoscientific witness that scours the victim's body for genetic traces of sexual assault. This genetic technoscientific witness has been inscribed with the presumed powers of DNA typing. Many police investigators, lawyers, and scientists described how the SAEK can now be trusted to produce unbiased and credible evidence of what occurred during a sexual assault and who the perpetrator was. "It is what it is ... The evidence will either corroborate it or not ... You just find what you find," a forensic scientist I interviewed said. Sexual assault investigations, according to one police officer, are now made abundantly easier when "there is solid DNA saying that it was done."

The reassembled SAEK enacts new meanings of victim's bodies. Sexually violated bodies become sites where identity codes of sexual assault perpetrators could be found. After DNA analysis was introduced, the SAEK's forensic script encouraged a victim to not only conceptualize her own body as a crime scene, as the previous iteration of the SAEK had, but also as a crime scene that held evidence that had the potential to protect herself and the public from a sexual predator (Mulla, 2014). Mulla's ethnographic research of sexual assault exams in Baltimore illustrates the expectations that now fall on victims to protect the DNA evidence on their body. These expectations have become embedded into the SAEK's forensic script.

Nurses and police investigators whom I interviewed underscored the ongoing risk of victims destroying DNA evidence by washing or wiping their bodies after a sexual assault. To avoid these risks, they actively encourage victims to abide by the SAEK's script. One nurse stated, "A lot of girls hate this; however we tell them don't wipe before and don't wipe after. Because the more they wipe, the more evidence is wiped away." A police investigator expressed a similar sentiment and said, "The average victim who is the true victim[24] will normally shower ... It's not good for me ...Unfortunately that will wash off most DNA." Many of the websites for SACTCs now contain instructions for victims on how to ensure that DNA evidence is preserved. With the increased sensitivity of DNA testing, as compared to forensic technologies of the past, the expectations on victims to demonstrate greater vigilance in abiding by the SAEK's forensic script have increased. Mulla's (2014) and Doe's (2012) empirical studies on victim's experiences with the SAEK suggest that this pressure weighs heavily on many victims of sexual assault. Victims are compelled to avoid showering or using the washroom after their assault until a forensic nurse or physician is available to collect the evidence. They are similarly compelled to consent to the forensic exam and make the DNA evidence on their body available to investigators and scientists, despite what may be their hesitations in undergoing an invasive medical exam after being sexually assaulted. In the popular accolades given to DNA evidence, these implications that DNA typing has for victims are rarely described.

Appearances of Stability

Since the heated controversies over DNA evidence in the 1990s, forensic DNA evidence has gained status as "a signifier of truth" (Jasanoff, 1998) and a "veritable truth machine" (Lynch et al., 2008) that is trusted to give credible and reliable evidence of criminal identities. New forensic science television shows, such as *C.S.I.* and *Law and Order: Special Victims Unit* reinforce and reproduce the apparent credibility and objectivity of DNA typing in the public imagination (Nelkin & Lindee, 2004). Controversies over DNA typing now appear closed and faded into the past, and the SAEK's network has been seemingly restabilized after years of controversy.

Police investigators, lawyers, and forensic scientists that I interviewed described forensic DNA evidence as infallible and powerful, as a "priceless" "silver bullet" that is "absolutely indispensible" and "the best crime fighting tool we've had." One police investigator explained the power of DNA evidence by saying, "It makes your evidence a certainty ... You can't fight the DNA." Many investigators suggested that DNA evidence has become a new standard in sexual assault investigations. One explained, "Without that evidence, our hands are tied. It goes from 100% we can ID the person to your word versus his." Another proclaimed that "society has become DNA driven, where it's almost made it a bit more difficult for our cases in court if there is no DNA."

Questions about the science behind DNA typing have all but disappeared in Canadian courtrooms. One defence lawyer described the prospect of challenging the science of DNA typing as "a tough row to hoe" and another admitted he has "never tried to fight it." A forensic scientist recalled the criminal trials of the 1990s and remembered how often he had to respond to the criticism that "DNA [was] junk science." In recent years, he said, he is rarely challenged on the scientific methods of DNA typing. Referring to the possibility of a defence challenging the science, he suggested:

> They can try. They are not going to get very far, there is just so much precedent that I can just [snap snap snap] you know, "Are you telling me, and this court, that the whole forensic community is misguided? The FBI, the

European countries, Australia, that we are all using DNA technology, and it is wrong?" It is a hard argument.

In the course of a mere two decades, DNA evidence has gone from a highly contested terrain to a sealed set of practices that many claim to be immune to criticism. The controversies around DNA evidence momentarily destabilized the SAEK's network when they called into question the purpose of the SAEK's evidence and how it should be made to speak in the laboratory. But, as consensus grew around DNA analysis as a trusted science and new technologies were put in place to coordinate its use, such as the DNA databank, the SAEK's network regained some stability. As DNA evidence has secured a largely accepted place in the SAEK's network, proponents of DNA typing have strengthened their claims about the technology's power to protect the public from crime. In 2006, the National DNA Databank wrote, "The power of DNA to help solve crime is undeniable, and continued support of the National DNA Databank will help to ensure that this power is harnessed and used effectively to support the safety of Canadians" (National DNA Databank, 2006, 7).[25] Two Canadian forensic scientists extended claims about DNA protecting public safety by projecting that if the National DNA Databank and DNA typing were used to their full capacity, more than 1800 sexual assaults would be prevented each year in Canada (Hepworth & McLeod, 2005). While it is unclear how these numbers were calculated, they reinforce the doctrine that the reassembled SAEK, and associated forensic technologies for DNA analysis, can protect the public, and mainly women and girls, from sexual assault.

Instabilities in a Stabilized Network

Networks can appear stable even when their components are not. Singleton (1998) argues that stabilized networks can contain sources of instability. Although DNA evidence is now widely celebrated as being the most powerful forensic tool for sexual assault investigations to date, questions are stirring about whether its powerful reach may have in fact gone too far. Technologies for DNA typing have become more precise since Jeffreys's first RFLP DNA typing technique. PCR and STR techniques have made it possible to generate DNA profiles from immensely small or degraded samples left on clothing or on bedding several years earlier. The result has been that both older and newer DNA profiles can be detected from a single item. This change, according to several police investigators, has introduced new challenges for sexual assault investigations. One officer in a police forensic identification unit explained that DNA testing was "almost to the point where it is getting too sensitive now. You can actually pull off very subtle DNA samples, you know that could be quite old. DNA in the right conditions will last a long time." He explained that a common consequence is that sexual assault cases often involve multiple DNA profiles from clothing in the SAEK, which can make investigations more time consuming and complex. He said, "We are getting a lot more of these multiple DNA hits ... If anything that can make our work a little harder."

Multiple DNA profiles found on clothing can raise new questions about a victim's sexual history. The forensic identification unit officer recalled one case where this was made most clear:

> We had a woman who was sexually assaulted and when her panties were sent away it came back with 5 male DNA profiles. It sounds terrible right? However ... they went back to the woman, they explained that they got these profiles and she was embarrassed and ashamed. This is what I would call your normal young woman, she's in her mid to late 20s ... One of the questions I asked the detectives to ask her is does she have what I referred as, her "going out" panties? You know? Nice ones that are sexy and she feels good in etcetera and she wears out for special occasions. And why we ask is that statistically studies have shown that semen is amazingly resilient. And you can wash those panties up to 12 times and semen will still be detectable with a DNA profile. So yah what can happen is in this case, it turned out that that's what these were. They were her "fancy panties." So like she said, they only got worn a couple times a year. So in all fairness, 5 DNA profiles sounds terrible, and you can only imagine how a defence would run with that ... However, when you think about it, these five profiles ... happened over a time period of about 3 years. Which you know ... in today's environment that is [a] perfectly acceptable kind of thing. So it is explainable. But that's a little caveat we have to watch because science has gotten so good at detecting DNA that you know you've got to be careful.

Despite legal restrictions on interrogating a sexual assault victim about her sexual history in the courtroom, there are no parallel restrictions for police investigations. DNA typing techniques have opened up new avenues for interrogating a victim's sexual history. In the quest to see the identity of a non-consensual sexual partner(s), DNA typing techniques can allow investigators to see *and count* DNA profiles of consensual partners as well. With DNA analysis, a victim's sexual past can be made visible and ripe for interrogation. When a victim's sexual past deviates from what investigators see as "normal," this can inspire new lines of questioning and become something that victims are asked to explain. Another investigator described the effect that this kind of interrogation about multiple DNA profiles can have:

> You ask "why are there three or four persons' DNA in these samples, in your underwear?' ... And this person might get the impression that you think they sleep around ... And that might make them even more reluc-tant to speak to you ... Now you are prying more and more into people's personal lives, bad enough to pry into something that has been forced on somebody, but now you are having to ask about something that might be perfectly consensual, they might be pretty reluctant to telling you when there is consensual sex too.

In some cases, DNA can inspire and seemingly demand answers to new questions about a victim's past that investigators might feel compelled to follow. The confusion, pain, and discomfort that can result for victims are new sources of uncertainty and instability in the SAEK's network that are often invisible in accounts that praise DNA's power.

The rise of DNA typing has also introduced new challenges for Canadian forensic labs and the scientists working in them with an unprecedented increase in the workload. Between 2000 and 2006, the requests that the CFS received for biological forensic testing increased by 224% (Auditor General of Ontario, 2007). The increase in workload has led to significant backlogs in forensic testing. In 2007, CFS's average turnaround time was seventy-three days (Auditor General of Ontario, 2009).[26] Around the same time, the RCMP's promised thirty-day turna-round for forensic testing was rarely met and was routinely extended to 108–20 days (Hepworth & McLeod, 2005).[27] In addition to the pres-sures from increased requests for DNA typing, two legislative changes significantly increased the workload of Canadian forensic laboratories. In 2008, Bill C-13 and Bill C-18 added 172 designated Criminal Code offences for which police could seek a warrant to obtain a DNA sample from a suspect. Many politicians framed Bills C-13 and C-18 as neces-sary advancements for public safety. Senator Marilyn Counsell claimed, "There is no doubt that Bill C-18 moves law, justice, and the safety of all Canadians forward" (Counsell, 2007, 6). However, for forensic labs that were already overburdened with increasing numbers of requests for DNA typing, these bills threatened to increase the existing backlogs.

In 2009, the Standing Committee on Public Safety and National Security cited the increasing weight that Bills C-13 and C-18 placed on forensic labs as a significant problem and declared that CFS and Quebec's provincial forensic laboratory[28] were in "emergency mode" (18). As a way of resisting the new workload, CFS refused work that arose from the legislative changes and sent it to the federal RCMP lab, which was under similar pressures (Standing Committee on Public Accounts, 2008). In 2009, CFS publically stated that an additional $11 million dollars over three years would be required to meet their increasingly large workload demands (National DNA Databank Advisory Committee, 2010). I found no evidence that this boost in funding was ever granted.[29] In the absence of increased government funding, forensic labs have responded to increasing workloads by turning to automation, which has introduced new complications.

Automation and robotics have been implemented in forensic laboratories to increase the speed and decrease the cost of forensic DNA analysis (Hudlow & Buoncristiani, 2012). In 2007, CFS reported that automation decreased their staffing costs in the biology unit by 42% (Auditor General of Ontario, 2007)[30] and by 2009 their turnaround time had been reduced to sixty-six days (Auditor General of Ontario, 2009). While these numbers were celebrated by CFS, they obscured the complexities behind automation. There are many steps in the DNA analysis process that cannot easily be automated. Forensic scientists Hepworth and McLeod (2005) suggest that automation can have the undesired effect of creating "bottlenecks" (4) at the non-automated stages of the process.

Automation, in some Canadian labs, has also produced undesirable results. The RCMP forensic lab introduced automation to DNA typing in 2005. Shortly after doing so, laboratory officials announced that automation had "increased casework capacity two or three fold while improving the timeliness of reporting results" (Auditor General of Canada, 2007, 14). However, internally, forensic scientists had been raising concerns that the automated systems were producing haphazard results. For a year, the scientists' concerns were not addressed and were dismissed with claims from the management that the automated system was in fact reliable. Only after the Auditor General made the scientists' concerns public in 2007 did the RCMP offer to retest all evidence that had been analysed with the automated process (RCMP offers, 2007).

In this case, automation increased the RCMP's workload and, in doing so, did not contribute to eliminating bottlenecks, backlogs, and delays.

Despite the mounting pressures on forensic labs and the evidence of backlogs and delays, some Canadian groups are lobbying for increased DNA testing. The focus has been most predominantly on expanding the DNA databank. Approximately 36,000 convicted offender profiles are uploaded to the DNA databank each year with court orders (Standing Committee on Public Safety and National Security, 2009). The committee that conducted the 2009 statutory review of the DNA Identification Act problematized the administrative time required for the Crown to secure court orders for DNA sampling and highlighted the regional differences in the number of court orders that are made (ibid.). The committee recommended that DNA samples be taken immediately upon conviction, without a court order. The committee estimated that the change in legislation would increase the number of profiles uploaded to the DNA databank to 113,000 per year, which, they argued, would justify a significant increase in the funding for the National DNA Databank. The Canadian Association of Chiefs of Police have gone further to propose that police should have the right to obtain DNA samples upon arrest for offences such as sexual assault, aggravated sexual assault, and murder (Canadian Association of Chiefs of Police, 2011). Several lawyers I interviewed suggested that the Canadian Charter of Rights and Freedoms would likely prevent any attempt to eliminate court orders for DNA sampling. However, as the history of DNA typing and banking in Canada has shown, it is imaginable that claims of collective safety could be used to surpass Charter rights. These changes could put significantly more pressure on the already taxed forensic labs.

Amidst assertions that DNA evidence has drastically improved sexual assault investigations, there continue to be voices of contention and sources of uncertainty and instability. While these tensions have yet to significantly rock the stability of DNA typing and the SAEK as a reliable genetic technoscientific witness of sexual assault, they may have potential to do so in years to come as pressures mount on forensic laboratories and critics of DNA's limitations in sexual assault investigations become more vocal.

The introduction of DNA typing significantly shifted the terrain around the SAEK. When DNA typing was introduced in 1989, the action around the SAEK moved from hospital emergency wards and SACTCs, where it had been in the 1980s, to scientific laboratories, legal courtrooms, and government boardrooms. With the development of new experts to administer the SAEK in hospitals in the 1980s and new expert laboratory practices to analyse it in the 1990s, the action around the SAEK was moving further from its origins in the rape crisis centres of the anti-rape movement.

DNA typing reassembled the SAEK into a genetic technoscientific witness of victims' bodies. This new witness could be *made to speak* with DNA analysis to reveal the identities of sexual assault perpetrators. This new witness, unlike the victims', whose credibility was routinely challenged in court, could be trusted to give reliable and credible testimonies about sexual assault and the identity of its perpetrators. Through DNA, the SAEK gained new stability as an indispensible forensic tool. However, despite the SAEK's stability, controversies around forensic DNA typing have not ended or faded into the past. Rather, some of the old controversies have re-emerged and new ones loom on the horizon.

5 Instability Within: The Technoscientific Witness in Contemporary Practice

When physical evidence runs counter to testimonial evidence, conclusions as to physical evidence must prevail. Physical evidence is that mute but eloquent manifestation of truth which rate high in our hierarchy of trustworthy evidence.

J. Baeza and B. Turvey (2002, 169)

The overriding purpose of forensic testing is not to collect evidence to catch the rapist but to validate a woman's claim that she has been raped ... Her story is not believed by investigating officers until a medical professional confirms it verbally and in writing ... If cuts, bruises, emotional trauma or, most important of all, rape sperm are not collected in the kit, the police are predisposed to believe that the woman is lying, that no crime has occurred.

J. Doe (2003, 305)

Since its development in late 1970s, the Ontario SAEK has acquired a central role in medicolegal responses to sexual assault. The SAEK, and the forensic exam it is part of, now shape how police, nurses, doctors, lawyers, and some victims respond to sexual assault. As the kit has become an integral part of medical and legal practice, consensus has seemingly grown about its capacity to produce reliable and credible evidence of sexual assault. The network of relations around the kit has seemingly stabilized, and as a result, the kit has gained stability as a veritable technoscientific witness of sexual assault. However, beneath these stable exteriors are new sources of instability. Lingering questions, deliberations, and debates have continued well into the new millennium about how the kit should be used, what its value is, and to whom. Its apparent stability is thus a complex one riddled with sources of instability, tension, and controversy.

This chapter examines contemporary medical and legal practices to reveal how the SAEK acts and is enacted as a technoscientific witness of sexual assault. It takes up the questions: Who benefits from the kit's stability as the technoscientific witness?, What purposes are served by its stability?, and What are the costs of its stability and to whom? Star (1991) proposes a critical shift in actor-network theory (ANT) when she argues, "It is both more analytically interesting, and politically just to begin with the question, *cui bono?* [to whose benefit?] than to begin with a celebration of the fact of human/non human mingling" (43, emphasis added). Beginning with the question of *cui bono*, for Star, repositions power and marginality as central concerns within studies of technology and the networks in which they are apart. Here I take up Star's reimagined ANT to ask of the contemporary SAEK: *cui bono?*[1]

Embedded in the question *who benefits from the SAEK* is a more theoretical one: for whom is the kit and its network stable? Star (1991) contends that "no networks are stabilized or standardized for everyone" (44). Instead, "a stabilized network is only stable for those who are members of a community of practice who form/use/maintain it" (44). Actors who are not part of a community of practice but who rely on stabilized technologies and networks, rarely feel their stability. Star argues that the public stability of technologies and networks is often accompanied by "the private suffering of those who are not standard – who must use the standard network, but who are also non-members of the community of practice" (43, emphasis added). For these non-members, Star argues, stabilized networks are not sources of order and stability as they are for members. Instead, stabilized networks are more often "source[s] of chaos and trouble" (42). There are many sources of chaos, trouble, and private suffering embedded in the SAEK's stabilized medicolegal network. This chapter explores these instabilities within the SAEK's network.

Tracing how the kit acts and is enacted as a technoscientific witness in practice allows for a more detailed picture of the SAEK and its network to emerge. Seeing the kit in practice reveals the part it plays in a heterogeneous, medicolegal network of actors. The SAEK is not simply an innocent technology, but is instead an actor playing a role in medical and legal practices that have the potential to create trouble and private suffering for some victims. The SAEK's politics as a technology born out of a distrust of victims of sexual assault become clear when we see the kit working alongside other actors responding to sexual assault. It becomes clear how the kit can act as a technoscientific witness for excluding, coercing, and interrogating some victims. Here, I argue that these multiple roles that the kit plays in medical and legal practice, and the private suffering and instability that result, are embedded in the medicolegal network's stability. They are crucial to how the kit acts and maintains stability as the technoscientific witness of sexual assault.

The contemporary SAEK acts within a radically different medico-legal network than the one it entered in 1979. It is within this significantly altered network of relations that the SAEK now travels through sexual assault exam rooms, police stations, forensic labs, and courtrooms. To understand how the kit acts and is enacted in medical and legal practice, it is necessary to first sketch this altered terrain that supports and shapes the flow of medicolegal action through which the contemporary SAEK moves.

The SAEK's Contemporary Network: Suspect Victims and Contested Terrains of Practice and Expertise

Victims reporting sexual assault to police face a different network of actors, practices, and protocols than they encountered in the 1970s. New actors and new power relations between them have been created with the development of specialized nurses and hospital centres for forensic exams, designated police units for sexual assault investigation, new protocols for police training in sexual assault and hospital protocols for evidence handling and collection, and more professionalized counselling services in rape crisis centres. These developments are often praised for vastly improving the sensitivity and efficiency of sexual assault services and increasing rates of reporting, prosecution, and conviction of sexual assault. However, these accolades obscure the suspicion of sexual assault reports that continue to pervade policing practice, and hide the brewing tensions within and between rape crisis centres and hospital-based sexual assault treatment centres. Although the SAEK's contemporary network is markedly different than it was when the tool was first announced in Ontario, controversies over practice and expertise endure, and doubts about victims' truthfulness persist.

Sexual assault cases continue to have dramatic attrition rates in the Canadian legal system. Of the small percentage of sexual assaults that are reported to police (less than 10%), only an estimated 0.3% of sexual assault cases result in conviction (Johnson, 2012). Long before reaching a courtroom, police investigators dismiss a significant number of sexual assault reports as false and unfounded – a term reserved for those cases where police conclude that no actual violation of the law took place or was attempted (Crew, 2012; Doolittle, 2017). Evidence suggests that long-standing societal myths about sexual assault filter into law and police practice – myths that only strangers commit sexual assault, victims always respond emotionally to sexual assault, victims are less credible if they are sexually active, and women routinely fabricate reports of sexual assault (Crew, 2012; Corrigan, 2013b; Hodgson, 2010; L'Heureux-Dubé, 2001; Mack, 1993). Many police continue to draw on these myths when assessing the truth of sexual assault reports (Dellinger Page, 2010; Flood & Pease, 2009; Johnson, 2012; Quinlan, 2016; Russell, 2010). Institutional-ized racism, sexism, classism, ableism, and transphobia continue to cast greater suspicion on women of colour, Indigenous women, disabled women, and transwomen who report sexual assault (Irving, 2008; Lie-vore, 2005; Martin, 2005; Mulla, 2014; Odette, 2012; Sallomi, 2014). Police investigations of sexual assault are often characterized by what Hodg-son (2010) calls a "systemic predisposition" (173) to disbelieve victims of sexual assault. Describing this further, Hodgson writes: "There is no other crime against the person where police so routinely, thoroughly, and explicitly question whether the crime occurred in the first place and attempt to discredit the complainant" (179).

Blair Crew's (2012) groundbreaking study on sexual assault polic-ing in Ontario revealed the significant rates of sexual assault cases that police deem to be unfounded. He discovered that in some police organziations in the province, 32.45% of sexual assault reports are dis-missed as unfounded, as compared to other crimes, where the highest unfounded rate is 3.43%. Unfounded rates from the 1970s indicate that little progress has been made in the past few decades to curb the alarm-ing rate at which police dismiss sexual assault reports. In 1975, Ontario police dismissed 40% of reported rapes as unfounded, as compared to 6.5% of assault reports, 8% of robbery reports, and 5.9% of breaking and entering reports, and laid charges in only 30% of rape cases (Police crime statistics, 1975). The consistent rates of police dismissing sexual assault reports suggest that police continue to cast far more suspicion on victims of sexual assault than they do on victims of any other crime.

Police investigative practices are often shaped by the anticipated needs of the prosecution in court. The investigative reports that police write, according to Mulla (2014), are made up of "anticipatory structures" (154) that reflect the expected steep requirements for a successful prosecution. The continued legal demand for corroborative evidence and the steadfast biases in court against sexual assault victims put pressure on police to only pursue those cases where there is ample corroborative evidence and victims who are more likely to appear credible in the eyes of the judge and jury. Cases for which there is forensic evidence from the SAEK are seen as more promising for successful prosecutions (Corrigan, 2013a). In this way, police investigators serving as the gatekeepers to the criminal justice system (Tasca, Rodriguez, Spohn, & Koss, 2012) are charged with the task of evaluating cases on the basis of the anticipated needs of a successful sexual assault trial.

Although sexual assault investigators now receive specialized training, myths about sexual assault victims being inherently untrustworthy prevail. Some of these myths are embedded within the training itself. Crew (2012) notes how police training manuals often contain assertive statements about the pervasive problem of false sexual assault reports. Baeza and Turvey's (2002) police training manual provides the "Baeza False Report Index," which lists sixteen "red flags" of false reports of sexual assault. "Red flags" include a range of indicators such as, a victim missing her curfew on the night of the assault, having a psychiatric history, having a past experience of reporting a sexual assault, and displaying "TV behaviour" (177).[2] These "red flags" help to construct a narrow conception of *real* victims as those who are untouched by men, unlabelled by psychiatrists, and seemingly unaltered by their own sexual assault. Training texts, such as Baeza and Turvey's, act on and within police practice by shaping and coordinating how sexual assault investigation is done and the attitudes and beliefs that police bring to their work.

Echoing Baeza and Turvey's index, several police investigators that I interviewed outlined the signs that raise doubt in their mind about the truthfulness of a sexual assault report. Some of these included a victim's emotional state, her level eye contact, the number of times she touches her face, changes in her tone, word choices, and the level of detail when she describes the assault. Other behaviours, motives, and narrative styles were also noted as suspicious. A few police officers indicated that they are most doubtful of a victim who cannot describe her assault backwards, does not know if her assailant was circumcised, reports with her boyfriend the next morning, or does not report at all, is desperate for attention, and has a motive such as she was late for curfew or she needed an excuse for an upcoming exam.[3] Rape crisis advocates reported that certain victims receive this scrutiny more than others. Women of colour, Indigenous women, disabled women, women with mental health diagnoses, transwomen, and sex working women are far more likely to be viewed with significant suspicion by police and have their reports labelled as unfounded. Despite the last four decades of changes in the medicolegal network, institutionalized prejudices, coupled with institutionalized myths about sexual assault, continue to bleed into many sexual assault investigations and influence the level of suspicion that falls on victims.

Against this backdrop, the contemporary SAEK works as a technoscientific witness that police can use to corroborate or challenge a victim's report of sexual assault. Police scepticism of sexual assault victims' reports bolsters the credibility and authority of the SAEK, and the evidence inside it, to confirm or disprove the sexual activity in question. Corrigan (2013b) argues that police distrust of sexual assault victims turns the forensic exam into a "trial by ordeal" (924), which places demands on sexual assault victims that far exceed those placed on victims of any other crime. In a criminal justice system that demands corroborative evidence of sexual assault, victims are expected to consent to the invasive forensic exam and comply with the kit's forensic script to protect the evidence on their bodies and facilitate its successful recovery (Mulla, 2014). Sexual assault kits, in Corrigan's view, are used as a test of not only the validity of the victims' report but also her commitment to the criminal justice process. The systemic suspicion of sexual assault victims heightens the stakes of the forensic exam and intensifies the expectations on victims to enter the kit's network.

While the enduring distrust of victims in police investigations and the criminal justice system more generally have remained relatively constant in the past few decades, other parts of the SAEK's contemporary network are in flux. It is this network of shifting relations that victims encounter when they report a sexual assault to the police, a hospital, or a rape crisis centre. These shifting relations are characterized by complicated power relations between advocates and medical professionals, and contemporary uncertainties and controversies over the politics and expertise of sexual assault treatment, advocacy, and care.

New claims of expertise in sexual assault care and advocacy are sparking new tensions between Rape Crisis Centres and Sexual Assault Care and Treatment Centres. Funding pressures and institutional constraints are leading many RCCs and SACTCs to renegotiate and redefine their identity, politics, and practices. All of this tension and renegotiation is occurring within a seemingly stabilized network that is dramatically different than it was forty years ago.

Contemporary anti-rape activism in Canada is operating in "a radically altered terrain" (Beres, Crow, & Gotell, 2009, 144) from its origins in the collectively run rape crisis centres of the 1970s. The pressures to professionalize sexual assault advocacy and transform RCC collectives into hierarchically run services in the 1980s and 1990s have coalesced with more recent pressures from severely diminished funding for activist organizations and demands from funders to redefine RCCs as depoliticized service providers for sexual assault (Beres, Crow, & Gotell, 2009; Doe, 2012; Russell, 2010). Many contemporary RCCs have moved away from feminist peer counselling to professionalized counselling by social workers and psychologists. New post-secondary programs, such as George Brown College's Assaulted Women's and Children's Counselor/Advocate Program, now offer formalized training for new advocates in RCCs. Most centres now operate under a hierarchical or modified collective model and few collective organizations remain. One advocate in a collectively run RCC in Ontario described the decline in collectives by saying, "It is slowly becoming more lonely ... We are trying to operate as a circle in a triangle world." Under these pressures, sexual assault advocacy is being redefined in some RCCs as a more individualized service that is managed and conducted by professionals.

In the face of the pressures to transform RCCs, one rape crisis advocate suggested that the broader anti-rape movement has become "much more mainstream ... We've lost some of our activism along the way." Sunny Marriner (2012), an activist and advocate in Ontario, explains this further: "In a culture of increasing credentialism and professionalization, we see sexual assault centres moving even further from the radical challenges of feminism, the expert knowledge it developed, and its at-once hopeful and skeptical aspiration to alter the social terrain for women" (412). This context has shaped the relationship that many RCCs now have with the feminist politics that sparked their opening. Beres, Crow, and Gotell's (2009) recent national survey of contemporary Canadian RCCs revealed that funding constraints on RCCs have significantly reduced their time for political activism, but have not significantly reduced their feminist politics. Marriner argues the opposite: "In what appears to be a bid to win the contest of 'experts' with psy- and institutionally based services, many of todays SACs have attached credibility to being perceived as 'every bit as professional' as government-created apolitical victim-service models" (445).

According to Marriner, this shift suggests that RCCs have been complicit in a larger effort to displace feminist expertise in psy- and institutionally based services. RCCs have, she argues, replaced their own feminist expertise with professionalized, apolitical, models of practice. Rose Corrigan's (2013a) detailed examination of rape crisis services in the United States draws a similar conclusion. She argues that the pressures on RCCs to work collaboratively with state agencies have significantly reduced their independence and autonomy. As a result, many RCCs adopt a professionalized, social service model in place of what was an explicitly feminist model. Explaining this trend, she writes, "Feminism ha[s] become an increasingly difficult and risky political position for advocates to maintain" (45).

While several of the advocates that I interviewed explicitly identified themselves and their centres as feminist, many others did not. The advocates who did not identify as feminists emphasized instead the strengths of the professionalized counselling and support services that their centres offer. In contrast to the early RCCs of the 1970s, which were often hubs of activism, radical feminist politics, and feminist expertise on rape, contemporary RCCs are renegotiating and redefining their identities and politics on a shifting terrain of insecure funding and pressures to professionalize, hierarchize, and de-politicize.

Accordingly, rape crisis advocates now play a different role in the SAEK's network. The level of involvement that advocates had in delivering training on the kit and consultations on its design and use in the early 1980s has significantly lessened. Advocates are now rarely involved in design consultations on the SAEK, while SANEs have primarily taken on this role. Advocates in RCCs now more often receive training on the kit from police and nurses than the reverse. Although advocates continue to support victims during the SAEK exam (when they are given permission to do so),[4] their relative positioning to the kit has changed. They have been pushed further into the periphery of the kit's network and have little influence on how the SAEK acts and is enacted in medicolegal practice.

These changes in RCCs have coincided with medical actors developing and asserting new claims to expertise over sexual assault treatment, advocacy, and care. SANEs are transforming terrains of expertise on sexual assault by increasingly taking up advocacy and public education practices that RCCs once claimed. Ontario's thirty-five SACTCs vary substantially in their histories, models of practice,[5] compositions of staff, and funding structures. Despite their differences, contemporary

SACTCs are commonly credited with providing supportive environments for victims, shortening SAEK exam times, improving collaboration with police, and producing higher prosecution rates (Ericksen, Dudley, McIntosh, Ritch, Shumay, & Simpson, 2002; Stermac & Stirpe, 2002). In chapter 3, I illustrated how the kit contributed to solidifying SANEs' expertise in sexual assault treatment. For much of the 1980s and 1990s, SANEs' expertise was focused on the SAEK exam and emergency medical care. Now, many contemporary SACTCs have expanded their services to include professionalized counselling. This counselling is incorporated into existing professionalized models of medical care, and as such it fits within the frame of professionalized sexual assault services, which funders and policymakers have promoted in recent years.

With the SACTC's range of individualized medical and counselling services, the centres have been dubbed a "one stop shop" (Burnett, 2007)[6] for sexual assault victims. Reflecting an outcome that some advocates feared when the first SACTCs opened in the 1990s, this new language for SACTCs renders RCCs and the advocacy they provide seemingly redundant and situates SACTC staff as central actors in the SAEK's network. Structural changes in contemporary RCCs and SACTCs, in which some Ontario RCCs are being subsumed or replaced by SACTCs, have made this shift most clear. In Sudbury, the Sexual Assault Crisis Centre closed and was replaced with a hospital-based sexual assault treatment centre that provides SAEK exams and counselling (Sudbury to get, 2011). When the new program was announced, some media reports downplayed its distinct purpose in comparison to its predecessor and cast the program as the "*new* Sexual Assault Crisis Centre" (ibid.). Some RCCs are thus being literally erased by SACTCs and RCC advocacy is being pushed further to the sidelines of medicolegal practices around sexual assault. Amidst these shifts, SANEs and SACTCs are claiming expertise in sexual assault, advocacy, and care. In so doing, these centres are adopting and transforming feminist-inspired anti-rape expertise, services, and practices that were once the sole purview of RCCs and the anti-rape movement.

Sexual assault care in the SACTC is shaped, and is at times determined by, the SAEK's forensic script and the medicolegal model of which it is part. Under the medicolegal model for sexual assault care in the SACTC, sexual assault nurses are placed in the contradictory position of being an objective advocate: s/he is caught between being a supportive medical caregiver and an objective forensic evidence collector

(Corrigan, 2013a; Doe, 2012; Du Mont & Parnis, 2003; Martin, 2005; Mulla, 2014). Sexual assault nurses, according to Doe (2012), are working within "a competition of cultures" (20) between law and medicine. Several nurses suggested that they deal with these competing pressures by siding with the SAEK's forensic script, which they argue is in the victim's best interests. One said, "It sounds very callous, but it's for their own protection … You have to tell them right off the bat … you don't want to hear all the extra details." Another said, "I don't need to know if it's not on the form." SACTCs, along with the SAEK, have contributed to transforming sexual assault care from a largely non-professionalized, community-based service in RCCs to a professionalized, medicalized service largely dictated by the needs of the legal system.

In addition to the legal demands on SACTCs, these centres are also under significant pressure to conform to the medicalized model of the hospitals in which they are situated, where care is defined as individualized intervention and service (Doe, 2012; Mulla, 2014). This has generated significant criticism from some advocates in rape crisis centres, who maintain that these pressures have heightened the medicalization of sexual assault care and have moved sexual-assault support services further from their origins in feminist social movements. One explained, "A lot of the treatment centres are not informed from a feminist perspective … They retain a lot of the values and myths from hospital and clinical practices that they were intended to eliminate." Another described the services in the SACTC in her community as "not client centered, certainly not empowerment based, and certainly not feminist." While a few advocates noted that the SACTCs in their communities were run by nurses and social workers who actively resist the pressures to adopt a medical model, these examples were often described as anomalies.

As the primary users of the SAEK, SANEs have secured a place in the contemporary medicolegal network as experts on sexual assault. Consequently, many SACTCs have expanded their scope to include community outreach and public education. Much of SACTC's public education work focuses on advertising SACTC services and educating women on the kit exam (Colby, 2008). According to SACTC network newsletters, however, some SACTCs have adopted a broader purview of public education, including the development of sex awareness curriculum for secondary schools (Kaplan, 2004), training sessions for lawyers, police, and paramedics (Furst, 2005), and education programs for secondary schools on the gendering of toys (Toppozini, Maxwell, & Mesch, 2003), drugs (Fitzgerald, 2006), personal safety, and bullying (Maxwell, 2006). Some descriptions of these public education programs describe sexual assault as a "personal safety issue" (Maxwell, 2006) that involves "three stages of healing: victim to survivor to 'thriver'" (Caufield, 2003). Sexual assault, in these descriptions, is defined narrowly as an individualized experience, which fits into medicalized definitions of trauma. Educational programs from this framework are a stark contrast to many anti-rape activists' education programs in the past, which began with the premise that sexual assault is not an individual problem, but instead is a social and political problem that requires a collective response to address the trauma that it produces.

In some communities, SACTCs' public education work has generated significant tensions with advocates in rape crisis centres. One advocate from Northern Ontario described SACTC's public education work on the SAEK in relation to the racism she sees permeating SACTCs and the medical system they are part of. She said,

> I couldn't emphasize enough the racialized experience in [*city name removed*] … It's kind of like we don't have missionaries anymore, we have *medical missionaries*. They are those kinds of people who think that Indigenous people are the fringes of a dying society, who should either just jump on board with mainstream culture or fuck off. (emphasis added)

SANEs and other experts from the SACTCs, she explained, often act as white "medical missionaries" who attempt to "educate" Indigenous women and women of colour on "healthy relationships," sexual assault, and the proper uses of the SAEK. She reflected on some of the SACTC public education programs and said, "There are huge gaps in the level of understanding that most practitioners have around the social determinants of health, and the context in which Indigenous people are living." Other advocates pointed to SANEs' increasing purview in the kit's network as an illustration of how hospitals are appropriating and transforming anti-rape advocacy and education. One advocate argued that SANE's educational work is failing to reflect the experiences of women of colour and Indigenous women and is abusing the expertise that SANEs appropriated from RCCs. These sentiments point to the growing tensions and perhaps a new contested ground between RCCs and SACTCs over expert knowledge on sexual assault.

Despite the expert status that SANEs and other professionals in SACTCs have secured over sexual assault care and advocacy, their positioning within the medical system is relatively tenuous. Corrigan (2013a) describes how the services that SANEs provide are grossly underfunded and "ghettoized" (130) in American hospitals. A similar situation exists in Canada. In 2008–9, the funding structure for Ontario SACTCs shifted from secure funding from the Ministry of Health and Long Term Care to global funding allocated by individual hospitals. One SANE described global funding as "very much stat driven" and allocated purely on the basis of the volume of patients that a hospital program serves. The global funding model prioritizes emergency wards that can serve up to 300 patients a day and other similar departments, and not SACTCs, which in some cases serve approximately 35–45[7] cases a month. Several nurses noted how the shift to global funding has resulted in significant cuts to nursing, counselling, and management staff in many SACTCs.[8] For some SACTCs, the cuts have eliminated available funds for nursing education and training. One SANE reported that her centre does not have the funds to pay nurses for their training time on revised SAEKs. She said, "The nurses aren't paid for their time to do it ... yet they have to maintain a high level of knowledge and competency especially in the forensic sides of things ... That is a real challenge." While significant resources are being funnelled into revising SAEKs, the actors using them are not given the resources necessary to learn the intricacies of the revised tool.[9]

In addition to funding challenges for SACTCs, SANE programs routinely struggle with retention and recruitment of registered nurses (RNs) (Louisa, 2010; Siedlikowski, 2004). Sexual assault nursing salaries are often unpredictable and variable, as many of the hours worked are on-call and with marginal compensation. To supplement the salary, many nurses seek more secure employment in other positions, which can interfere with on-call hours. Beyond the inadequate salary, nurses commonly identify burnout, compassion fatigue, and vicarious trauma as significant challenges and barriers to SANE work (Dempsey, 2009; Sievers & Stinson, 2002). A nurse explained the emotional difficulties of the work and the increased feelings of vulnerability to sexual assault that she and her female colleagues experience by saying, "We see in here all these things that we are at risk for." Retention difficulties along with cuts in funding have resulted in the understaffing of many SACTCs.

In the face of pressures to cut costs and increase numbers of sexual assault nurses, some centres have expanded their recruitment to include

registered practical nurses (RPNs).[10] In 2006, the RPN scope of practice was expanded by the Ontario College of Nurses, allowing RPNs
to conduct most steps in the SAEK exam.[11] Following this change, the
number of RPNs in sexual assault nursing has significantly increased:
in one northern SACTC, four out of ten nurses are RPNs. Including
RPNs in sexual assault nursing is a significant shift from the early
years of the SAEK, when physicians were considered to be the only
qualified experts that could conduct the SAEK exam. In the 1980s and
1990s, when physicians began to resist the SAEK exam as an undesirable responsibility, RNs replaced them and became expert SAEK users.
Now, as RNs are becoming increasingly difficult to retain, particularly
in northern Ontario, RPNs are increasingly filling the void. Similarly
to how RNs solidified and secured their expertise over the SAEK in
the 1990s, medical actors are now describing RPNs as "professionals"
who have the "knowledge, skill, and judgment required for the role"
(Fitzgerald & Rioch, 2009, 4) of the SAEK examiner. The designated
expert user of the kit has shifted in response to increasing pressures to
cut healthcare costs, expanding scopes of practice for medical practitioners, and doctors' and nurses' resistance to administering the SAEK.

Several nurses I interviewed expressed concern about the future of
SACTCs. With the cuts to funding, the difficulties in recruiting and
retaining RNs, and a predicted rise in retirements in the near future,
these nurses suggested that new challenges and changes in SACTCs
are likely on the horizon. SANEs' positioning in the SAEK's network is
thus not fixed, nor is the network itself inherently stable. Rather, within
its stability lies much instability; the contemporary SAEK's network is
characterized by shifting practices and relations between actors. How
does the SAEK act in this shifting network? For what purposes? For
whom? What sources of instability lie within the SAEK's stability as a
technoscientific witness? And, who feels these instabilities? It is to these
questions that I now turn.

The SAEK: *Cui Bono?*

Within the SAEK's changed medicolegal network, the kit is often
praised for enhancing sexual assault investigations, increasing conviction rates, and improving victims' experiences in the criminal justice
system. Many police, nurses, and forensic scientists that I interviewed
claimed that the SAEK has "advanced a lot of investigations," "led
to better convictions," and "brought the justice system a little further

ahead." Others maintained that the SAEK is "beneficial for victims" because it often "give[s] legitimacy" to their stories and, in so doing, "assist[s] complainants in allegations of sexual assault." A few rape crisis advocates agreed and asserted that the SAEK "has been helpful" for victims and has assisted "women [in] build[ing] their cases." In one of the most emphatic descriptions of the SAEK, a police officer proposed that it was in fact necessary for justice: "It is so important that she takes the kit, if she doesn't take the kit like, I mean if you don't have that in court, we are not going to do her justice."

In addition to being a potential tool for justice, many police and some forensic nurses also described the kit as a tool that can empower victims of sexual assault and grant them agency and control. Police officers and administrators said that victims can choose to use the kit for their benefit, and some forensic nurses portrayed the kit and the forensic exam as a victim's personal choice. Several nurses stressed that in the SACTC, "everything is [the victim's] choice." In the "one-stop shop" (Burnett, 2007) of the SACTC, nurses present victims with what one described as "a big menu ... We just say, here's what we have to offer, take whatever will be of use to you, and leave the rest." Many nurses and police suggested that giving a victim the choice of having a SAEK exam helps to counteract the lack of autonomy and agency that characterize sexually violent attacks. "We want to let them take some of that control back," one nurse explained. In these descriptions, the kit is configured as a stabilized tool that victims can use to regain control, autonomy, and choice.

These portrayals of the kit suggest a solidified consensus around its value. However, alongside common depictions of the kit as a valuable and empowering tool, several police, nurses, and lawyers stated that, in practice, the kit does very little to definitively prove that a sexual assault occurred and rarely facilitates convictions. Many advocates in rape crisis centres asserted that, in practice, the SAEK rarely works to benefit victims. These advocates described the kit's questionable value. One stated, "In theory it's awesome, but in practice, and legally it doesn't work for the survivor," while another stated that the kit's "intention is great, but it doesn't really serve its intention." Reservations about the kit's value in practice are reflected in the scholarly literature on sexual assault investigation and prosecution (Du Mont & White, 2007). Although some studies have found that evidence from SAEKs increase the likelihood of arrest and conviction (Campbell, Patterson, & Bybee, 2012; Campbell, Patterson, Bybee, & Dworkin, 2009), others have found that kit evidence increases the likelihood of a police charge and arrest, but not conviction (Johnson, Peterson, Sommers, & Baskin, 2012), and others still have found that kit evidence has minimal, if any, impact on outcomes of sexual assault investigations and trials (Feldberg, 1997; McGregor, Du Mont, & Myhr, 2002; Sommers & Baskin, 2011).[12] The contradictory evidence and assertions about the SAEK's value reveals some of the uncertainties embedded in the SAEK's seemingly stable network.

Scholarly accounts of the wide range of experiences that victims have with the SAEK exam paint a similarly complex picture. Some victims report that the forensic exam was an empowering and affirming experience, whereas others say that it was retraumatizing and revictimizing and akin to the sexual assault itself (Doe, 2012; Du Mont, White, & McGregor, 2009; Mulla, 2014). One survivor[13] in this study recollected the support and encouragement she received from the nurses and police after she reported her sexual assault. She described her SAEK exam as "very empowering." In contrast, another survivor said that her exam felt "just like the sexual assault … People are doing things to your body and you don't even know what they are going to do next … You had no choice in the process over your own body." The variability in victims' experiences of the kit is rarely acknowledged in discussions about the successes of the SAEK's contemporary network.

By framing the SAEK as an inherently empowering choice for victims, some police and nurses gloss over the practices that limit and impede victims' choice and control in the forensic exam. They shield from view the ways that medical actors have placed the SAEK out of reach for many victims and how police and lawyers have used it to coerce and interrogate some victims in sexual assault investigations and prosecutions. More specifically, they hide the multiple roles that the technoscientific witness can take on in practice, and the instabilities that arise from them. Mol's (2002) discussion of multiplicity provides some useful tools for understanding how the kit adopts multiple identities in practice. Mol proposes that in medical practice, diseased bodies become *multiple*. In her ethnography of a Dutch hospital, she examines hospital sites where diseased bodies are enacted in different ways. In the physician's office, she argues, a diseased body is enacted as pain and other visible symptoms, whereas in the laboratory, it is enacted as diseased cells under a microscope. Through these different enactments, Mol contends, the diseased body becomes multiple. The SAEK is similarly multiple in practice. While the SAEK might be a tool that aids investigation and prosecution of sexual offenders in some cases, in others it is enacted as a tool for the exclusion, coercion, and interrogation of victims.[14] The technoscientific witness becomes multiple as it takes on these other identities in practice. The kit's multiple identities create instabilities in the SAEK's seemingly stable network that create trouble and private suffering for some victims.

When the SAEK travels through its network, it links practices in the examination room, the police station, the forensic laboratory, and the courtroom, and acts with other medicolegal actors to assemble legal "truths" about sexual assault. The following reveals how the SAEK acts as a technoscientific witness that can be used to exclude, coerce, and interrogate some victims in three points of action[15] in the SAEK's network: (1) accessing the SAEK, (2) staging the SAEK, and (3) testing truth with the SAEK. By tracing the multiple identities that the kit can adopt in these sites and the practices that complicate victims' choices within them, I do not mean to deny victims' capacity to make choices in regards to the kit or to erase the range of more positive experiences that some victims have with it. Instead, my interest here is in the different practices that impede a victim's choice and control and create instabilities within the medicolegal network.

A Tool for Exclusion: Accessing the Kit

Praises for the SAEK as a potentially empowering technology for sexual assault victims obscure the challenges that many victims face accessing the kit.[16] Despite the fact that Ontario SAEK exams fall under publicly funded health care, kits are not equally accessible to all victims. Inadequate staffing and resourcing of SACTCs can significantly reduce victims' access to SAEK exams. A victim's geographical location, residency status, and financial means can also influence her access to the kit. For those victims who struggle to access the kit, it is not a stabilized tool that is empowering and affirming. Instead, it is a tool that can create significant obstacles and, in some cases, further pain and suffering. The SAEK's network is a standardized system that excludes many victims. Star (1991) writes, "There are always misfits between standardized ... technological systems and the needs of individuals" (36). Who are the victims whose needs do not easily fit within the bounds of the SAEK's network? And, what are the medical and legal practices that enact the kit as a tool for exclusion?

A victim's location and access to safe transportation can severely restrict her access the kit exam. The thirty-five SACTCs in Ontario that provide forensic exams are all located in urban hospitals and are responsible for providing services for large geographical areas that include both urban and rural communities for populations ranging from 11,000 to 1,000,000 people (Macdonald & Norris, 2010). Many advocates and nurses in this study reported that since these centres were established in the 1990s, hospital emergency wards are now more likely to refuse victims care. General emergency wards now commonly redirect victims to the nearest sexual assault treatment centre. While specialized expertise around sexual assault has in some ways improved services for victims, this has come at the cost of reducing the number of professionals who are willing and trained to do forensic exams. Victims' access to the forensic exam has been accordingly reduced.

General emergency wards now more commonly redirect victims to the nearest SACTC, which in some regions can be several hours away. The consequences of this can be significant, particularly for victims who do not have the financial capital, capacity, or desire to travel after a sexual assault. In 2011, three women who had been sexually assaulted were turned away from the Ottawa Hospital because there was no SANE available and no general emergency nurse or doctor willing to conduct the SAEK exam (Seymour, 2011). The emergency room staff reportedly told the victims to go to a treatment centre in a neighbouring city, which was over an hour away. One woman was unable to finance the travel required and was forced to relinquish the possibility of a kit exam. Months later, when her assailant was brought to trial, the judge dismissed the charges, concluding that it was "disturbing" that no SAEK exam had been done and that without it, her evidence was too unreliable to find the accused guilty (Seymour, 2011, C1).

Advocates, nurses, and police working in communities in Northern Ontario described how victims often face even greater obstacles accessing the SAEK in northern regions of the province. Hospitals in small northern communities are usually more willing to conduct sexual assault kit exams than hospitals in urban regions of the province. However, many nurses and police working in northern regions of the province suggested that victims in northern communities are encouraged to travel to specialized sexual assault treatment centres for the forensic exam so that they can receive more comprehensive services. Depending on where the victim lives, getting to a specialized treatment centre in Sioux Lookout, Kenora, or Thunder Bay can involve several hours of travel by car or, in some instances, by airplane.[17] To fly by airplane, victims must finance air travel themselves, apply for air travel funding from their employer, or, if they live on a reserve and are eligible, obtain funding from their band council. Receiving air travel funding usually requires victims providing a detailed rationale and documentation justifying the need for travel, which can place significant pressures on victims to disclose that they have been assaulted. Despite the evidence that treatment centres have improved sexual assault services for victims (Hatmaker, Pinholster, & Saye, 2002; Sampsel, Szobota, Joyce, Graham & Pickett, 2009), they have had the unintended consequence of limiting victims' access to the kit. These treatment centres have transformed the SAEK into a tool that is primarily reserved for victims who have the financial means and access to secure safe transportation, as well as the time and desire to travel potentially long distances immediately following a sexually violent attack.

Advocates outlined how victims' residency status, ability, and past experiences with institutionalized racism can also influence their capacity to access the kit. One advocate described how far out of reach the SAEK's network is for victims with no legal status in Canada: "Going to the hospital wasn't something that they could do without an Ontario Health Insurance Policy card, and even if they were able to get to the hospital and have the kit done, going to the police was not an option." Another advocate noted that kits are often similarly out of reach for victims who have intellectual disabilities and are unable to "make the call to say 'this has happened to me, what do I do?'" She explained,

> If they have intellectual disabilities or developmental disabilities, they might not be able to advocate for themselves in the same way … This is a whole population who are probably not getting the rape kit done because they don't have the information and the capabilities to even get themselves there or get themselves connected to someone who could give them that information.

Picking up a similar theme, another advocate described how victims' past interactions with the medical system, particularly those who have been diagnosed with psychiatric disabilities, can impact their ability to access the kit. She stated, "The more marginalized you are, the more chance you've also been in contact with this [medical] institution." She continued with a rhetorical question: "If you have disabilities, or you have mental illness, or you are a young woman who's … [been] medicated or diagnosed, are you going to the same profession to try to be believed?"

Several other advocates described the impact that victims' past experiences of institutionalized racism can have on their ability and desire to access kit exams in hospital-based treatment centres.[18] One advocate working in an RCC in Northern Ontario described the impact that racism in the medical system has on victims' access to kits:

> It [is] white faces in the hospital. And most of women coming through the door are Indigenous … A lot of the hospitals in the North are notorious for poor treatment of Indigenous people … So it's not like you are going to walk through the door and see yet another white person in this hospital system and feel immediately safe. That doesn't happen. So many people just get so tired of struggling with the mainstream organizations, they just don't bother.

Many of these advocates reported that victims who face these types of barriers accessing the SAEK often decide not to have a forensic exam and sometimes do not even seek medical care after their assault. For these victims, the kit's network can be enacted as a site of exclusion, and the kit an inaccessible technology.

Barriers to accessing the SAEK also play out in the context of the exam. Part of the kit's protocol involves police taking, and often not returning, pieces of a victim's clothing for laboratory analysis. For victims living in poverty, this can be a difficult sacrifice. Nurses and rape crisis advocates recalled instances where victims were pressured to give up their clothing despite the fact that they did not feel they could afford to do so. One advocate in Southern Ontario said, "I remember distinctly one woman saying, 'Do you know how much jeans cost? I can't afford that!'" A nurse in Northern Ontario remembered a debate she had with a police officer who wanted to take a child victim's only winter coat away for analysis. Despite the winter conditions outside and the mother's adamant concerns about how she would not be able to afford another coat, the police officer argued that the coat was necessary for the investigation because it was crucial to "setting the scene" of the sexual assault. By stipulating that victims must sacrifice pieces of their clothing in order to receive a full forensic exam, the SAEK excludes some victims from fully accessing its network. Only those victims who can afford, and are willing, to sacrifice their clothing in the name of forensic science can receive a complete SAEK exam. In this way, the kit can become a technology that creates and maintains barriers to criminal justice processes.

Within these medical and legal practices that restrict victims' access to the SAEK, the kit takes on a new identity. It can become a technology that is largely reserved for a particular type of victim – an ideal implicated user[19] of the kit. This ideal user is a victim who has the means and available finances to secure safe transportation to a SACTC, the time and desire to travel to an SACTC immediately following a sexually violent attack, as well as the residency status, language skills, and confidence in the medical system that are all necessary to access the kit. Since the medicolegal practices are geared towards supporting this type of victim, she may feel the stability of the kit and its network. However, for the victim whose needs are different than the ideal implicated user's, the SAEK has little stability or benefit. In practice, the kit can thus become a potentially inaccessible tool that works to exclude some victims from participating in its network. It becomes a tool that has potential to cause more trouble and private suffering for some victims than healing or empowerment. Other identities of the kit arise when police present the SAEK to victims, and when police and lawyers use the kit's contents in sexual assault investigations and criminal trials.

A Tool for Coercion: Staging the SAEK

Victims who are able to access the SAEK face various pressures to comply with its forensic script. Police investigators' suspicion of sexual assault victims, and faith in the kit's capacity to enhance sexual assault investigations and improve conviction rates, increases the weight that they place on SAEK evidence and leads some to pressure victims to consent to the SAEK exam. Chandler (2010) suggests that technologies can erode choice and autonomy when they are considered to be indispensible to action and are configured as "offers you cannot refuse" (15). Tracing how police stage the SAEK to victims reveals how, in practice, the kit is often turned into an offer that victims cannot easily refuse. More specifically, it illustrates how the kit can become a tool for pressuring and coercing victims into complying with criminal justice processes.

Sexual assault victims' decisions on whether to proceed with criminal justice procedures are not purely expressions of personal volition (Kerstetter & Van Winkle, 1990). In the context in which forensic evidence in the kit is highly valued in police investigations and prosecutions, a victims' decision to consent to the SAEK and other criminal justice procedures is far more complex than just an individual choice (Mulla, 2014). Kerstetter and Van Winkle's (1990) early study on sexual assault victim decision making pointed to the many pressures on victims to pursue criminal prosecution. More specifically, they drew attention to the pressure that police place on victims to comply with criminal justice procedures, particularly those victims whom police believed would be more credible witnesses in court. The SAEK can become involved in police investigators' pressure on victims to comply with criminal justice processes when police first present the SAEK exam to victims.

When a victim reports a recent sexual assault to police, investigators present victims with the possibility of having a forensic exam. Often in the back of a police cruiser on the way to the hospital, police investigators stage the SAEK by describing its components and the significance of kit evidence to the police investigation. In Ontario, as in other provinces in Canada, victims of sexual assault must give their written consent before having a kit exam. The majority of police investigators that I interviewed said they do not pressure victims to consent to the kit exam and always give victims the opportunity to make an informed decision about the SAEK exam. However, when some police described *how* they present the kit to victims, it became clear that victims are not always given the freedom to make independent and informed choices about the kit. In fact, police practices suggest that it is more likely that victims are pressured and potentially coerced into having a kit exam done. In these instances, the kit becomes a tool that erodes choice and demands victims' compliance.

Many police stage the SAEK as a necessary tool for achieving justice and preventing perpetrators from victimizing others. Some do this in explicit ways, using the language of justice and crime prevention when they describe the kit's importance to victims, while others convey these sentiments more implicitly. One nurse described a tense encounter she observed between a police investigator and a victim, where the investigator was stressing the importance of the kit: 'There was a young woman who was in here who was sexually assaulted and the police kept saying 'you've got to do a rape kit, you've just got to do a rape kit!!' And she said, 'Why do I have to? I don't want to. I know him, I know who did this to me, why do I have to do a kit?' And the police kept saying, 'Well, we need the evidence!'" For this police investigator, the kit and its evidence were essential to the investigation, whereas, for the victim, both were superfluous. And yet, in this interaction, the investigator transformed the kit into a seemingly compulsory tool that would make the police investigation possible.

Not all police staging practices are as direct. One police officer claimed, "I've never had a person outright refuse a kit. I've had them say no, and I'll just sit them in the car, and I'll drive them there myself, and I'll say 'Well we are here now, let's go in and do it.'" The drive that the police investigator referred to was to a hospital SACTC a half an hour away. He continued by saying, "They are usually pretty up for anything as far as we *need* to get this done" (emphasis added). By driving victims to the hospital without their explicit consent, this investigator's actions enact the kit as a seemingly inevitable component of the criminal investigation. In these instances, the kit is no longer an option, but instead an expectation.

In contrast to both of these examples, several police investigators said that they consistently tell victims that the SAEK is "voluntary" and that they "never insist on doing something medically to her that she doesn't want." However, many investigators also claimed that they commonly tell victims that the SAEK will ensure "a more thorough investigation" and will give police "a better shot at finding who did this, so that he won't do this to you or anybody else again." Another said that he routinely describes the SAEK as a step that a victim can take to "help us out," and in return, police can "giv[e] them some sort of justice," as though one seemingly depended on the other. Victims who have been coerced into going to a hospital treatment centre and have been told that the kit is a necessary step for ensuring the safety of themselves and others have little room to choose *not* to have a kit exam.

SANEs often have different staging practices for the SAEK. Many SANEs I interviewed stressed that they routinely tell victims that the kit exam is not compulsory, and instead present it as one of the many options at the SACTCs. However, several advocates in rape crisis centres provided a different view of nursing practice. These advocates described instances they have observed where SANEs were indirectly shaping and controlling victims' choices around the kit. SANEs constrain victims' choices around the kit, they asserted, by not providing victims with complete information on how the kit may positively or negatively affect the sexual assault investigation and trial.[20] In addition, they said that some SANEs approach victims with the expectation that they are having a kit exam. One advocate described this further: "I think there is a tendency within the sexual assault treatment program to say, 'Women who have been sexually assaulted in the following ways *should* have the following exams, and *should* have the following medical treatment' and there is a tendency to sort of give the choice, but it is sort of *a prescribed choice*" (emphasis added). Corrigan (2013b) argues that SANE programs are built on the assumption that victims want to report and participate in the prosecution of their sexual assault. In this context, she argues, post-rape *medical* care has come to be defined as a *forensic* process. Against this backdrop, the forensic exam becomes an expected action as some SANEs, like police, stage the SAEK in ways that reinforce the kit's forensic script and, in so doing, constrain victims' choices in the forensic exam.

Police and nurses' SAEK staging practices reinforce notions of the ideal implicated user of the SAEK who follows the kit's forensic script and participates in forensic evidence collection to protect herself and others from violence. One survivor I interviewed recalled how much pressure she felt to follow the kit's script during her own experience of the exam; "All I was thinking was 'Be the best possible patient, be the best possible victim, answer all the questions.'" Two advocates described how often they have seen victims struggle under this kind of pressure. They repeated some of the things victims say when contemplating whether to consent to the forensic exam: "They say, 'Well, I need to go get the evidence, like I'm not a *good citizen* if I don't do that and get this guy off the street'" and 'Am I being a *bad victim* if I don't want to do this?'" (emphasis added). The SAEK and its presumed value as a tool for protecting people from violence puts pressure on victims to comply with its requirements, a pressure that is reinforced by police and nurses who stage the SAEK exam as a necessary or expected action for victims.

The direct and indirect pressures that some police and nurses put on victims are not equally applied. Advocates contended that police are far *less* likely to pressure women of colour, Indigenous women, disabled women, women in the sex trade, and women experiencing poverty to consent to the SAEK. Correspondingly, advocates said that white, middle-class women are far more likely to be seen as "undeserving" victims and more likely to be pressured to have the forensic exam. One described this by saying,

> I mean police discriminate, period. You could have ten women who all go to the police on the same day to the same officer who say, 'I've been sexually assaulted' and give the exact same story, but one woman is a disabled woman, one woman is under eighteen and living on the street, another

is a prostitute, another woman is from an upstanding family in [name of wealthy neighbourhood] and you would have ten different stories from the police: one is dismissed, one is laughed at, and one is carted off to the hospital for the rape kit.

Other advocates spoke more broadly about racism, ageism, and classism in contemporary medical and legal practice and described how these prejudices alter how some police and hospital staff treat victims. One said, "Depending on how the survivor acts, depending on the colour of their skin, depending on their age, depending on their perceived class, they are treated differently." While some victims are strongly encouraged and perhaps coerced into the exam, others are not. For these victims, the kit can be made inaccessible by police investigators' lack of faith in their reports of sexual assault. When the SAEK is staged differently for different victims, victims are pushed to either adhere to the SAEK's forensic script or accept that the script was not designed for them.

Despite common assertions that victims have complete autonomy around the SAEK exam, medical and legal practices around the kit suggest otherwise. These practices reveal that the kit can be used to coerce some victims into complying with police investigations and to dismiss others. In this context, choice carries little meaning and empowerment even less.

Obtaining (Un)informed Consent

A victim signing the consent form[21] in the SAEK is meant to symbolize her assent to and full understanding of the forensic exam and its consequences. Despite the existence of the SAEK consent form, there are unanswered questions about the extent to which victims are able to give full and informed consent to the kit exam. Some victims have past experiences with the kit, well-informed advocates, access to resources, and other supports that facilitate their informed consent to the exam. However, those that do not have these supports can face medicolegal practices that limit, and in some cases preclude, their informed consent to the use of the SAEK.

Much of the existing scholarly research on consent in the SAEK exam suggests that many victims do not feel that they consented to it. Du Mont, White, and McGregor (2009) found that many women did not know they had the option to give their consent. Doe (2012) similarly found that the assaulted women she interviewed had one of three experiences of "consenting" to the exam: "a) they had no memory of consent, b) they felt coerced into agreeing, or c) they believed their consent was necessary for the state to pursue criminal charges or otherwise 'protect' them" (15). Of the twelve women that Doe interviewed, none were informed about how the evidence would be used after it was collected or the possibility that it could be used against her during the investigation or the trial. The two survivors that I interviewed had similar experiences. One said, "It felt more immediate, like this is what we are going to do, and this is what happens ... It wasn't necessarily saying 'This is how the evidence is going to be used, or not going to be used.' They didn't go into that kind of detail ... I don't feel like they gave a big back-story to the implications of what can happen with the evidence or how it can be used or not used." Another survivor said, "It wasn't explained what was going to happen ... I have no memory of that ... I didn't know what it entailed, I didn't know what it meant."

The SAEK protocols stipulate that informed consent means that "patients [victims] must be able to understand the information that is relevant to making a decision about the use of the kit and be able to appreciate the reasonably foreseeable consequences of a decision or lack of decision" (emphasis added).[22] This definition mirrors the definition of informed consent in the Health Care Consent Act (1996) in Ontario, which states that consent is only considered informed if patients are given an explanation of a medical treatment's benefits, risks, side-effects, and alternatives, and the consequences of not having the treatment. Despite these protocols, victims' reports suggest that police and nurses have a significant amount of discretion when deciding what information about the SAEK is "relevant" for victims.

Several nurses and police said that they rarely explain the full consequences of the kit exam and how the evidence may be used in the investigation and trial. One nurse said that instead she explains the kit by describing the steps it involves "in the simplest terms as possible" so as not to overwhelm victims with too much information. A police investigator said: "People who are sexually assaulted, they are in terrible terrible turmoil and ... when something really awful happens to you, you can't go to the logical side of the brain, you are thinking emotionally, you can't reason, you can't make sense of things, and you can't remember things. And we know that." The assumption that people in trauma are unable to think logically has pervaded the handling of the SAEK since its origins,[23] and has historically been used to rationalize the absence of detailed information available to victims on how the kit acts beyond the exam room. While these practices may stem from an intention to provide more accessible and sensitive care for victims, they also dangerously reinforce notions of victims as traumatized individuals who do not want, or cannot handle, a more detailed explanation of the SAEK and its implications. Without being given information about benefits, risks, side-effects, and alternatives, and the consequences of not having the SAEK, victims are not well positioned to give their informed consent to the exam. The result is that the kit can become a technology that is even more difficult for victims to refuse.

A Tool for Interrogation: Testing Truth

After a SAEK exam has been conducted and the kit submitted for analysis, police investigators, Crown prosecutors, and defence lawyers interpret and use its analysed contents. Despite the fact that in Canadian sexual assault law, a victim's sexual assault report no longer requires corroboration from a witness or independent evidence, investigators and defence lawyers that I interviewed described how they routinely rely on the kit evidence to corroborate or challenge victims' reports of sexual assault.[24] In these practices, the kit becomes enacted as a techno-scientific witness of sexual assault, which police and defence lawyers can use to interrogate, challenge, and test the truth of victims' sexual assault reports.

Alongside police and defence lawyers' practice of using the kit as a tool for interrogation in investigations and criminal trials, there are ongoing controversies about what meaning can and should be derived from the evidence in the kit. Within and outside of Canadian courtrooms, medical and legal actors are asking questions about the relevance and meaning of the DNA profiles and evidence of injury inside the kit. Wrapped up in these ongoing debates are larger questions about who has the appropriate expertise to make truth claims based on the SAEK's contents and whether sexual assault is in fact visible on sexually violated bodies, and therefore whether its traces can actually be *seen* in the SAEK. What is visible in the kit is a reflection of the tools and techniques for analysis that medicolegal actors use to analyse its contents. What is made visible in the SAEK shifts with changes in technologies, legal definitions, and medicolegal practices. Visibility is an outcome of action and not an inherent state. In the ongoing controversies about what meaning should be made of the SAEK's contents, the value of the tool itself has been called into question. The instabilities arising from these ongoing controversies are embedded in the stability of the SAEK's network and the tool itself.

Making Meaning of the SAEK's Contents

How the physical injuries documented in the SAEK (or lack thereof) can be interpreted is highly contested. Recent victimization surveys and a significant body of contemporary medical and social science literature concurs that physical injuries from sexual assaults are rare (Biggs, Stermac, & Divinsky, 1998; Brennan & Taylor-Butts, 2008; Du Mont & White, 2007; Ledray, 2001; White & Du Mont, 2009). In 2008, Statistics Canada reported that 77% of sexual assaults involve no visible signs of physical injury (Brennan & Taylor-Butts, 2008). These findings have, however, had little influence on the SAEK's design. The SAEK's instructions still require SANEs to document traces of visible injury with photography and its body maps. To comply with the kit's demands for visible evidence of trauma, nurses may also use several tools to see traces of forceful sexual activity that might not be otherwise visible, including the colposcope, a large binocular microscope that nurses use to magnify and photograph micro-trauma in and around a woman's vaginal cavity, and toluidine blue, a liquid dye nurses use to determine the extent of cellular damage in and around a woman's vagina (White & Du Mont, 2009). Even though the kit remains built on an expectation of visible injury, several nurses that I interviewed indicated that visible injuries are in fact rare. And yet, visible injuries in SAEKs, they claimed, are often still taken to be necessary indicators of lack of consent in sexual assault investigations and criminal trials.

The majority of interviewed police officers stated that one way they determine consent is to look in the SAEK for evidence of trauma, bruises, torn clothing, broken buttons, skin under a victim's fingernails, and "redness, soreness, or any marks on the vagina." One police officer described visible injuries as "good solid medical evidence that will support the Crown's case when [the defence] are arguing consent." Crown lawyers described the common expectation among judges and juries of visible injuries on sexual assault victims. Defence lawyers also noted how visible injuries are often useful for Crown attorneys wanting to illustrate non-consent. The continued expectation of visible signs of force as signs of non-consent suggests that the pervasive legal

Figure 5.1 SAEK Physical Examination Form (Female), 2013 (© Queen's Printer for Ontario, 2013. Reproduced with permission. The materials are current to 2016 and are subject to change.)

PHYSICAL EXAMINATION FORM

Kit No. _____

Mark all injuries relevant to the assault as well as areas of tenderness and Alternative Light Source (Polilight/Woods light) findings on the diagram. Describe colour, appearance and size of injuries. Provide a brief history of injuries. USE QUOTATION MARKS IF YOU ARE USING THE EXACT WORDS OF THE PATIENT.

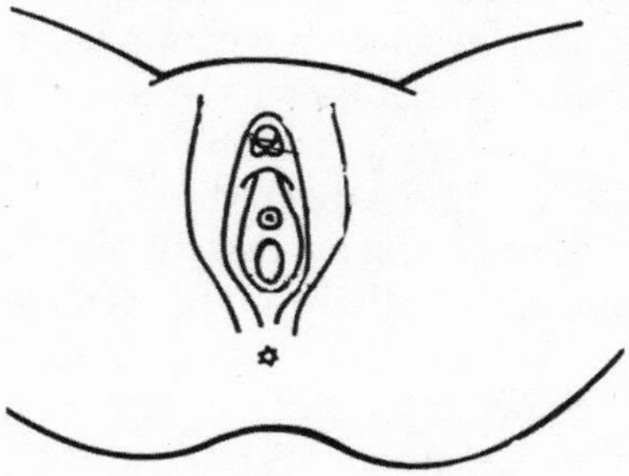

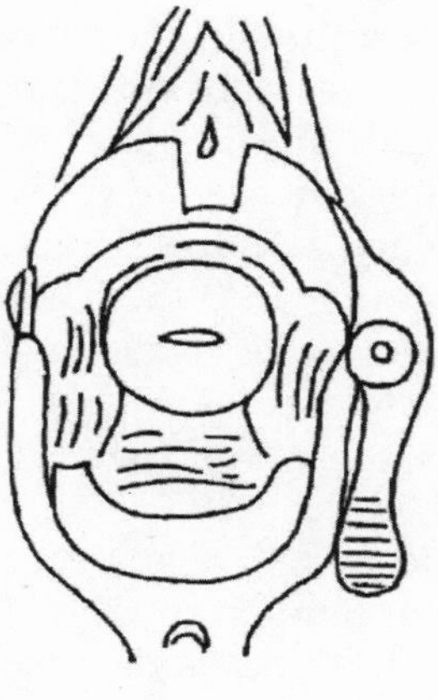

FOR FEMALE PATIENT

Labia Majora and Minora:

Posterior Fourchette and Introitus:

Vagina:

Cervix:

Anus and Rectum:

Os:

Left vaginal wall:

Right vaginal wall:

Discharge:

Physician/Nurse Examiner's Signature	Date	Time
CFS SAEK 2012	Hospital Records - White Copy	Police - Yellow Copy

requirement for evidence of force and violence in the 1970s and 1980s continues to pervade contemporary practice.

SANEs have increasingly been called as expert witnesses to testify that consent to a sexual act cannot be assumed from a lack of visible injuries. Through their testimonies, current medical understandings of how sexual assault appears (or does not appear) on violated bodies have moved into the courtroom for debate. Several defence lawyers described how they challenge this expert testimony. They contest SANEs' capacity to give expert evidence on injuries by arguing that SANEs do not have experience or training in examining women who have consented to the sexual activity, and therefore have no valid basis of comparison to the non-consenting bodies that they examine. In R. v. Thomas (2006), defence lawyers laid out a clear attack on the testifying SANE's expertise, and in so doing, successfully discredited the existing medical knowledge around injuries in sexual assault by proposing that "there is no science that can deal with whether the injuries are the product of consensual sex or not."

Defence lawyers have used the claim that there is "no science behind" the meaning of injuries in sexual assault to argue that the presence of injuries does not necessarily indicate a lack of consent. One defence lawyer said that he commonly argues in court that bruising and tearing in a woman's genital area is "fully consistent with vigorous or clumsy intercourse or foreplay" and is therefore meaningless evidence of lack of consent. Another defence lawyer similarly described how she discredits evidence of vaginal redness and soreness by asking women on the stand questions such as "Where are you in your cycle? If you are far from ovulation, do you have dryness issues? Is that customary? Do you normally have to use a lubricant?" Advocates described how many police mirror these interrogation tactics and often discount any signs of violence in the SAEK as indications of sexual activity or bodily functions. One said, "We've heard horror stories ... everything from 'she inserted her tampon roughly' [to] 'the bleeding was probably her period' ... Everything can get explained and the reason it gets explained is because we don't believe women in the first place." A survivor recollected how police discounted her own injuries:

Later on in the investigation I was being told that it wasn't going to go any further. I was kind of really upset because I wasn't feeling like I was believed. I wasn't feeling like I could provide any more evidence. I said,

"Well what about the tear? I don't understand. Is that not evidence?" And the investigating officer's response was "Oh you could have liked it like that" meaning, I could have liked it rough. So that's that. Basically it's fruitless evidence ... In cases of sexual assault, evidence doesn't actually matter.

Whereas some victims are discredited and disbelieved because they do not have injuries, others are discounted because they do. The ongoing debates about the meaning of victims' visible injuries or lack thereof are raising new questions about the worth of the kit itself. Either SANEs are arguing in the courtroom that the SAEK is unable to capture signs of non-consent or investigators and defence lawyers are dismissing the evidence that the SAEK provides. These are not the only sites of controversy where the kit's value is under question.

DNA profiles are another way that the contents of the SAEK become visible in police investigations and criminal trials. Despite the excitement around DNA evidence as a revolutionary technology for sexual assault investigation, DNA is not routinely found in SAEKs' contents. In one study at a Canadian laboratory, DNA profiles were retrieved from only 32% of kits (Gingras et al., 2009).[25] Making DNA visible can be hindered by decay in samples, testing procedures, and delayed collection of bodily evidence. In addition, if a perpetrator uses a condom or he does not ejaculate during the assault, DNA traces are less likely to be found. Despite the difficulties in making DNA evidence visible, its weight in the courtroom still looms large. Crown and defence lawyers discussed how useful the *lack of* DNA evidence can be for the defence. One defence lawyer summarized this by saying: "Non-evidence becomes evidence." Another explained,

We live in an era now where everybody has watched all these CSI programs and stuff, jurors in particular expect scientific evidence ... If it comes back with no DNA, for instance, I find that is helpful, because it is like the science processes didn't work, and they haven't come back with anything, that's almost worse than not doing any science at all.

The prevailing view that DNA analysis is an objective, fail-proof science supports the inference that the lack of visible DNA indicates that a sexual assault did not happen. Another defence lawyer summarized, "From my standpoint the best possible result is we did a sexual assault kit and we didn't find anything."

When DNA is visible, many police, nurses, and lawyers that I inter-viewed agreed that DNA is rarely useful in legally demonstrating that a sexual assault occurred. While there is consensus that DNA evidence can identify *who* was involved in a sexual act, most medical and legal actors emphasized that DNA evidence does not reveal *how* a sexual act was committed and, in particular, whether the act was consensual or not. Despite the ever-popular claims about the importance of DNA evi-dence in sexual assault cases, most lawyers said that DNA evidence from a SAEK is rarely relevant evidence in the courtroom. Because the vast majority of sexual assaults are committed by someone a victim knows, sexual assault trials more commonly revolve around questions of consent than of a perpetrator's identity (Brennan & Taylor-Butts, 2008; Statistics Canada, 2009). In these cases, DNA evidence that identi-fies a sexual offender is largely immaterial.[26]

The SAEK's inability to reveal consent or lack of consent to a sexual act has led many people to question its worth for victims. Many police, lawyers, and advocates criticized the kit by suggesting that when a per-petrator uses the consent defence, the SAEK and its contents become, in the words of one lawyer, a "moot point" that offers little insight into whether a sexual assault occurred. One defence lawyer questioned the kit's value to investigators and Crown prosecutors: "So if I'm right that the ... primary benefit of the sex assault kit is to [identify] the perpetra-tor, because most of them are between people they know, then *where's the value*? Because even if there are injuries, minor injuries, like nothing major, but if there is some minor redness, that doesn't mean they didn't consent" (emphasis added). The defence lawyer's question "Where's the value?" was answered by several investigators and defence law-yers, who claimed that the kit is in fact valuable to them, despite its inability to reveal consent, and has a contested capacity to make sexual assault traces visible.

A Valued Technoscientific Witness

Many police investigators and defence lawyers described the SAEK as a valuable tool for corroborating or challenging a victim's report of sexual assault. For these actors, the kit is a useful technoscientific wit-ness that can be employed to test the credibility of a victim's report. Many Canadian scholars have argued that the SAEK is primarily used to corroborate a victim's story of sexual assault (Doe, 2003; Feld-berg, 1997; Parnis & Du Mont, 1999, 2006), and several American and

British scholars have gone further to argue that police and defence lawyers often use the kit to challenge and discredit sexual assault victims (Corrigan, 2013b; Bumiller, 2008; Rees, 2012). While the kit's stability as a reliable technoscientific tool for corroboration benefits investigators and defence lawyers wanting to test a victim's report of sexual assault, it rarely benefits victims themselves.

In police investigations, the SAEK often takes on the identity of an objective tool for testing truth. Several police investigators suggested that the kit reveals the accuracy of victims' sexual assault reports. Significant contradictions between the SAEK's contents and a victim's report, they claimed, are often indications that the victim's report is false. Police training manuals, such as Baeza and Turvey's (2002), instruct police to use forensic evidence in this way and assume its objective accuracy when questioning the validity of a victim's sexual assault report. Baeza and Turvey's chapter on false reports opens with the quote "When physical evidence runs counter to testimonial evidence, conclusions as to physical evidence must prevail. Physical evidence is that *mute but eloquent manifestation of truth*, which rate[s] high in our hierarchy of trustworthy evidence" (169, emphasis added). Echoing this mantra, one police officer said of the SAEK, "When you know the truth, you know the truth. It's there in black and white. So that's what I use the kit for." She continued by saying, "We need corroborating evidence, you can't just come into the police station and say that you've been raped by your boyfriend last night and you want him charged. It's really not that easy." In police practice, the kit can shed discrediting light on victims' reports of sexual assault and lead some police to dismiss a report as false. When stacked beside a victim's uncorroborated report, evidence from the SAEK often has far more authority within police investigations. In this context, the kit becomes a tool that is assumed to have the power to reveal the accuracy and validity of victims' reports and can shape the direction of police investigations.

Not all of the SAEK's contents are seen as equally as important in police investigations. McMillan and White's (2015) study of police perceptions of forensic exams and forensic evidence revealed that many Ontario police investigators interviewed appeared to have a relatively narrow understanding forensic evidence, and see the primary purpose of the forensic exam as a means to collect DNA evidence to identify the accused. Over thirty years after the rape law reform in Canada that eliminated the legal requirement for corroborative, independent evidence identifying the accused, police continue to expect this form of

evidence in sexual assault investigations and view the SAEK as a necessary technology that supplies it. With the increasing popularity of DNA
evidence, police investigative practices now revolve more predominantly around DNA evidence collection and, accordingly, sexual assault
investigations around questions of identity (Sallomi, 2014). Despite the
fact that DNA evidence says nothing about whether a sexual encounter
was consensual, which is often the question upon which sexual assault
trials hinge, many police continue to treat DNA evidence as crucial for
investigations of sexual assault. One police investigator I interviewed
described the significance of DNA evidence in determining the likelihood of an investigation by saying, "Once we have DNA to back up one
of their versions, that's usually when we can proceed with one of the
charges." Another claimed that the emphasis on DNA evidence in the
criminal justice system puts pressure on police to ensure that they build
their cases with DNA evidence, regardless of its inability to speak to the
issue of consent. He said, "Society has become DNA driven, where it's
almost made it a bit more difficult for our cases in court if there is no
DNA." When DNA evidence identifying the accused is not available,
either because the victim did not consent to the exam or no DNA evidence was found on her body, her sexual assault report often becomes
more suspect in the eyes of police.

Police use the SAEK to know and test victims' truths in other ways
as well. Some investigators claimed that a victim's decision not to
have a SAEK exam is often a sign that the victim is lying about the
sexual assault and can be good justification for dismissing a report as
unfounded or false. Corrigan's (2013b) notion of the forensic exam as a
"trial by ordeal" (924) points to how police investigators use the exam
to test victims' sincerity and commitment to the criminal justice process. Victims in the United States who are willing to submit to the kit
exam, she argues, are more likely to be believed by police. Confirming
this trend, Alderden and Ullman's (2012) quantitative findings showed
that the odds of an arrest in sexual assault cases decrease by 57% when
a victim refuses to have a forensic exam. Illustrating how Canadian
police can transform the kit exam into "trial by ordeal," one survivor,
who was quoted in Du Mont, White, and McGregor's (2009), described
the purpose of the kit in police practice: "A kit was done so they could
find out if you were lying. I thought it was like a proof kit so that ...
the police had proof that the woman was actually telling the truth ... I
never thought it was for evidence ... against the perpetrator ... against
the person who has done the crime" (777). In line with these empirical

findings, some investigators I interviewed made sweeping generalizations and asserted that "real victims" want to have SAEK exams and will do anything to facilitate police investigations, whereas false complainants more regularly shy away from the SAEK exam.[27] Going further, one investigator claimed that the SAEK helps him detect and deter false reports of sexual assault. He explained,

> Sometimes there's complainants that come in and maybe the details are a little sketchy and you are not sure if this is going to be legitimate or not and you ask them to have the kit done and the prospect of that in a false allegation is too big of a step for someone to follow through on and they withdraw their complaint. So there's definitely going to be the odd case where you are stopping an unfounded allegation at the onset, once they realize what is involved.

The SAEK, in the eyes of some police investigators, is so closely coupled with truth that it can threaten potential false complainants. When police use the SAEK as a "trial by ordeal" or a "proof kit" for challenging victims' reports and threatening victims they deem to be suspect, they create instabilities in the kit's network for some victims. They turn the kit into a source of disorder and trouble for those victims seeking to be believed by the investigators they turn to for help.

The SAEK can take on a similar role in the courtroom. Many prosecution and defence lawyers stated, more often than not, that the DNA evidence inside the SAEK is irrelevant in the courtroom. They suggested that most sexual assault trials do not revolve around questions about the perpetrator's identity, but, instead, whether or not the sexual act was consensual. In these cases, DNA evidence identifying a sexual offender is often irrelevant. It is the other evidence from the SAEK – such as the forensic nurse's report of a victim's injuries and intoxication levels – that becomes more fertile ground for courtroom discussion.

Many Crown and defence lawyers explained that SAEK evidence is often more useful to lawyers defending accused sexual offenders than to lawyers prosecuting them. One prosecutor expressed frustration with how often the kit evidence routinely "backfires" on her in court. She said, "It's always been fodder for cross-examination and rarely would I ever tender it for my benefit." For her, the kit in court becomes an unpredictable actor that can work alongside defence lawyers to discredit victims and their reports of sexual assault. Along similar lines,

a defence lawyer said that she routinely relies on kit evidence to challenge the victim's testimony in court. Speaking very candidly, this lawyer said, "The kit is more useful to me than it is to the Crown."

Defence lawyers described how they use SAEK evidence to challenge
victims' testimony and credibility in court. Many said they examine
the kit evidence with a sharp eye for any inconsistencies between the
kit's reports and victims' police statements and court testimonies. Finding an inconsistency, no matter how slight, they explained, can throw
the victim's credibility into question in the courtroom and, alongside
it, her description of the assault.[28] One defence lawyer asserted that a
victim's level of intoxication that the SAEK documents is "usually a big
area that is ripe for cross-examination. Because she will say 'Oh no I
was terribly intoxicated,' 'Really, well you told the nurse you had two
drinks; how intoxicated were you?'" Another claimed that small details
in the nurses' notes on how the victim described her own injuries can
be useful in the cross-examination of forensic nurses: "[Beside] one of
the marks [on the body map, it] states, 'May have occurred tonight.'
So I'm able to say, 'That's the word that she used? *May*?" (emphasis
added). Defence lawyers also use the nurses' reports of bodily injuries
in the SAEK to note inconsistencies between a victim's description of
the perpetrator's level of force and the visible marks noted on her body.
One lawyer said,

> If the narrative that she tells [is] "I struggled, he had me down, and then
> he bit me, then he used his knees to pry my thighs,' you would expect
> there to be some kind of marks or bruises that go along with that narrative.
> So in that case, you would say things like "None of those things are there,
> so when he bit you, there is no mark, so he didn't bite you, you are not
> being truthful. It was [with] consent."

Rees (2011) and others have shown that SAEK evidence can serve as
a "mine of information for the defence team" (110), most particularly
nurses' notes about a victim's medical information and sexual history (Brown, Burman, & Jamieson, 1993; Kelly, Temkin, & Griffiths,
2006; Temkin, 1998, 2005). Despite the legal restrictions that have been
placed on admitting evidence of a victim's sexual history in court, Rees
points out that notes in the kit about a victim's sexual history continue
to inspire inferences and lines of questioning from the defence that
reinforce the assumption that victims with a sexual past consent to all
sexual activity. These defence strategies enact the SAEK as a tool that

works with defence lawyers to interrogate and contest victims' credibility and testimonies in court.

Given the potential ways that the SAEK evidence can act against victims in the courtroom, one defence lawyer asserted that, for victims, the SAEK evidence is simply not worth the discomfort that the exam causes – a claim that shakes to the core common depictions of the SAEK as a beneficial choice for victims. Stressing this point even further, she gave future victims the following advice on the SAEK: "Unless it is a stranger, you know, you were the one walking down the street and someone jumped out of the bushes, my personal opinion is, I would be telling somebody *not* to do it." Coming from a legal actor who claims to gain professional successes from a victim's choice to have a forensic exam, her advice to victims to not consent to the kit marks a strong indictment of it. It reveals how the SAEK's stability as a credible forensic technology works for some members of medical and legal communities around the SAEK, but more often than not, fails to work for victims. When the kit is used as a tool for interrogation, it takes on an identity that is in stark contrast to the common depiction of the SAEK as an efficacious tool for victims. The instabilities that result from these contradictions within the SAEK's network do not prevent the kit from being used, but rather exist as ongoing tensions within the kit's seemingly stable network.

Conclusion

The SAEK acts and is enacted in medicolegal practice that is geared towards testing the accuracy of a victim's reports of sexual assault. Despite continued controversies, the SAEK has maintained its status as the technoscientific witness of sexual assault. The apparent stability of the kit and its network benefit police and defence lawyers investigating the veracity of victims' reports of sexual assault. This stability has, however, come at a cost. It often works against those victims who do not fit the mould of the ideal implicated user of the SAEK or do not follow its forensic script by not preserving their bodily evidence or choosing not to have a SAEK exam. For these victims, the kit and its network are not stable, but are instead more likely to be a source of disorder, trouble, and private suffering. In contrast to common claims to the contrary, the SAEK rarely acts as an inherently empowering technology for victims. Instead, in many cases, the kit takes on more harmful identities and becomes a technology working alongside human actors to exclude, coerce, and interrogate some victims. These multiple identities of the kit and the private suffering that they cause are the hidden instabilities within the SAEK's network.

Instabilities do not necessarily threaten the stability of a network (Singleton, 1998). Singleton argues that "an established stabilized practice does not require stable and unequivocal entities as its constituents" (91); rather, she argues, instability can lie within stabilized networks and be part of what constitutes them. Instabilities in the SAEK's network are rarely made visible to those outside of the network. It is *because* public discussions about the kit rarely feature questions about victims' unequal access to the SAEK, the pressures that police put on victims to consent to the exam, or the ways that defence lawyers use the kit against victims in the courtroom that the SAEK can maintain its stability as a necessary technoscientific witness for sexual assault convictions and prosecutions. The erasure of instabilities within the SAEK's network thus contributes to its overall stability. Victims' private suffering is thus an essential (although largely invisible) part of the SAEK's contemporary network that has become central to how the contemporary kit acts and is enacted as the technoscientific witness of sexual assault in medicolegal practice.

6 Reassembling Technoscience: Troubled Pasts and Imagined Futures

In 2009, students, scholars, activists, lawyers, and survivors/victims met in Ottawa to celebrate and commemorate the ten-year anniversary of Jane Doe's legal victory. Jane Doe was raped in her Toronto apartment in 1986 by a serial rapist who had long been known to police as the Balcony Rapist.[1] Doe's attacker had raped four other women in her neighbourhood in the eight months before she was raped. Each rape was the same. The rapist broke into the women's apartments through their balcony doors and raped them at knifepoint. The Toronto Metropolitan Police saw the similarities between the four rapes and concluded that one man was responsible. Forensic evidence had, however, produced no investigative leads on the rapist's identity. With hopes of apprehending him, police watched women in Jane Doe's neighbourhood who they believed were likely targets and waited for his next attack. Despite their anticipation that he would rape again, the police did not publicize any information on when, where, and how the so-called Balcony Rapist was targeting and attacking women in Jane Doe's neighbourhood. The police later rationalized their silence by saying that a public warning might cause the women in the area to become hysterical and drive the rapist to a new area of town.

After Jane Doe was assaulted, she discovered that the police had known about the Balcony Rapist and had anticipated her rape. Angered that the police were wilfully putting women at risk, Doe and her feminist allies postered her neighbourhood to warn other women. "Rapist in this area ... Women be aware, be angry," the posters read. Within twenty-four hours of the posters going up, the Balcony Rapist's parole officer turned him in and he confessed to five rapes in Jane Doe's neighbourhood. In a historic legal victory twelve years later, Jane Doe successfully sued the Toronto Metropolitan Police for their negligence in failing to warn her that she was a potential target of a serial rapist and for violating her right to security and equality under the Charter of Rights and Freedoms through a police investigation that was shown to be marred by sexual discrimination.

The optimism that Jane Doe's victory had bestowed on activists, scholars, survivors/victims, and their allies was evident at the 2009 conference, "Sex Assault Law, Practice, and Activism in a Post-Jane Doe Era." Speakers reflected on the significance that this legal win had as the first in Canada in which police were held liable for failing to warn potential victims of crime and for sexual discrimination in policing practice.[2] In her keynote address, Jane Doe described the many problems that continue to plague contemporary policing of sexual assault. Little progress has been made, she said. Aiming a pointed critique at the SAEK, Jane Doe argued that the kit perpetuates dated rape myths, is seldom used in court, and rarely works in women's interests. And yet, she argued, it has become the new standard for sexual assault cases. This paradox, she contended, calls for activists' and advocates' renewed attention and action.

In the question period, a woman from the audience identified herself as one of the activists who was involved in advocating for the first SAEK. She underscored that in its early years, the kit was celebrated in feminist anti-rape communities as a success of feminist activism. With anger and dismay, she concluded her narrative about the kit by saying, *"They turned it against us."*

Almost forty years after the SAEK was first imagined in Ontario, it has secured status as a trusted forensic tool for witnessing traces of rape on victims' bodies – the so-called "crime scenes" of sexual assault. There is now an elaborate network of experts, technologies, and practices that surround, support, and work alongside the kit. The kit and its network are being credited by many with improving victims' experiences in hospitals, and for increasing arrest and conviction rates of perpetrators of sexual assault (Aiken & Speck, 1995; Campbell, Patterson, & Bybee, 2012; Campbell, Patterson, Bybee, & Dworkin, 2009; Campbell, Patterson, & Fehler-Cabral, 2010; Campbell, Patterson, & Lichty, 2005; Cornell, 1998; Little, 2001; Stermac & Stirpe, 2002).

This praise of sexual assault kits has been featured in the recent news media about the underfunding of forensic laboratories and lack of forensic nurses in rural areas in Canada, and the hundreds of thousands of untested sexual assault kits in the United States. An estimated 400,000 untested kits are currently being housed in police storage facilities and forensic laboratories in the United States, in large part because of the lack of resources for forensic analysis across the country and police investigators' continued practices of discounting and dismissing sexual assault reports (Sallomi, 2014). In response to these backlogs of untested kits, a growing number of legislators, advocates, and activists in the United States are publicly proclaiming the importance of forensic sexual assault kits for investigating and prosecuting sexual assault and promoting healing among victims. The United States Department of Justice, Office on Violence Against Women (2013) recently praised the powers of forensic exams in its *National Protocol for Sexual Assault Medical Forensic Examinations*:

> A timely, high-quality medical forensic examination can potentially validate and address sexual assault patients' concerns, *minimize the trauma they may experience*, and *promote their healing*. At the same time, it can increase the likelihood that evidence collected will aid in criminal case investigation, resulting in *perpetrators being held accountable* and further sexual violence *prevented*. (4, emphasis added)

Many American victim advocacy groups have chimed in with similar assertions about the powers of sexual assault kits to promote victims' healing, ensure justice, and prevent sexual violence (e.g., Human Rights Watch, 2009; Joyful Heart Foundation, 2015; National Center for Victims of Crime, 2016; Rape, Abuse, and Incest National Network, 2016). The Joyful Heart Foundation, a national advocacy group driving the "End the Backlog" campaign, uses the slogan "Hundreds of thousands of rape kits sit untested: each one represents a lost opportunity for healing and justice." In a strategic effort to draw attention to the consequences of kit backlogs on victims, sexual assault kits are being framed as necessary tools for victims' empowerment and access to justice. Some of the American media coverage of the sexual assault kit backlog has spilled over into Canada, where conversations about forensic laboratory funding and sexual assault services are gaining momentum.

The kit and sexual assault forensics have been similarly framed in Canadian news media reports about the perils of closing federal forensic laboratories and the lack of forensic nurses to conduct sexual assault kit exams in rural areas (Baziuk, 2011; CBC, 2012; CBC, 2015; Kane, 2015; Stone, 2014). Sexual assault kits and forensic exams have been cast as important for a victim's healing and useful for prosecuting and convicting sexual offenders. On a recent CBC radio broadcast, Sarah Tofte lauded sexual assault kits' abilities to "affirm the presence of the suspect ... affirm the victim's version of events ... and potentially match suspect[s] to other rape cases" (CBC, 2013). Sexual assault kits, she argued, are crucial for sexual assault investigations and prosecutions. Through all of this media attention on forensic sexual assault kits on both sides of the Canadian/United States border, sexual assault kits and the medical and legal practices they are part of have been cast and seemingly stabilized as effective means of putting sexual offenders behind bars and empowering victims. Obscured in these accounts, however, are the ways that the kit and its network often fail to live up to these accolades.

Despite the past few decades of legal reform, rapid changes in forensic technology, and developments in professionalized expertise and training in sexual assault forensics and treatment, many of the same issues that plagued rape cases in the early 1970s persist. Victims still report feeling dismissed, doubted, and re-traumatized by the legal system that claims to protect them (Maier, 2008; Mulla, 2014; Patterson, Greeson, & Campbell, 2009; Ullman, 1996, 2010). Fewer than one in ten victims of sexual assault in Canada report their attacks to police (Brennan & Taylor-Butts, 2008). Long-standing sexist and racist myths about sexual assault and sexual assault victims continue to filter into police practice and cast doubt on many women and men who report sexual assault to police (Irving, 2008; Lievore, 2005; Martin, 2005; Mulla, 2014; Odette, 2012). And still, only a fraction of sexual assault cases make it to the courtroom, and even fewer result in conviction (Johnson, 2012). The picture these data paint suggests that little has changed since feminist anti-rape activists began fighting for reforms in the 1970s that they hoped would transform legal responses to rape.

Bumiller (2008) and Corrigan (2013a) similarly argue that despite the institutional reforms it inspired, the anti-rape movement did not enact the widespread social change it intended to generate. Both scholars contend that the movement in the United States relied too heavily on the state and law to address sexual violence. The movement was co-opted to align with the interests of the neoliberal state, Bumiller argues, and demobilized by its alliance with the priorities of law and criminal justice, according to Corrigan. Corrigan and, to a lesser extent, Bumiller trace how anti-rape organizations in the United States were transformed and largely depoliticized by their alliances with law and the state.

A similar story is revealed in Canada's history of the SAEK. Anti-rape activists pushed actors in law, medicine, and science to improve medical and legal responses to rape and sexual assault. To do so, many activists had to conform their language and practices to fit within frames set by government funders, hospitals, and the criminal justice system. With the shift towards the professionalization of sexual assault care and the rise of medical forensic expertise on sexual assault in the late 1980s and 1990s, anti-rape activists were slowly pushed further into the margins of medicolegal practice. All of this occurred in a larger context of the rise of neoliberalism in the 1980s, in which the values of individualism and privatization took hold and social service agencies of all kinds were brought under the regulatory apparatuses of the state and forced to adhere to state-imposed policies and procedures (Bumiller, 2008). With the introduction of government funding, the advocacy and political work of the rape crisis centres in the 1970s became subject to state control and surveillance in the 1980s and 1990s. Through stringent funding requirements, government actors could regulate RCCs' political work and force them to work more cooperatively with medical and legal institutions. RCCs were pushed into a uniquely marginalized role in the growing network of medical and legal experts in sexual assault investigation, treatment, and care, which hampered their ability to oppose and influence institutional practice.

It was through these shifting dynamics that the anti-rape movement was transformed from a grass-roots social movement into a social service sector that now largely centres on a professionalized service-based model for post–sexual assault counselling and advocacy that in many ways facilitates, not resists, practices within the medicolegal system. In the early beginnings of the movement, anti-rape activists were lone voices pushing medical and legal systems to recognize rape as a pervasive social problem. They fought with institutions to make room for rape crisis advocates in hospitals, police stations, and courtrooms, and to address the systemic sexism that they saw shaping law and medicine. Forty years later, community-based advocates have been largely pushed to the sidelines of a network of professionalized medical and legal experts who offer institutionalized care and services for victims of sexual assault. One advocate reflected on this history:

> It's very interesting to see the same systems that didn't care at the beginning suddenly take on a very well-funded role in responding to these issues. It scares me that so many of them don't know their own

history. They don't seem to know that it was not social workers, that it was not nurses, that it was not doctors who understood the magnitude of these concerns. They were the very people who were brushing them off. It was people themselves, women themselves, who found their voices and fought.

Now, in the wake of several decades of institutional and legal reform, the network of professionalized services and expertise around sexual assault has obscured advocates' and anti-rape activists' history of transforming rape into a recognized social problem and fighting for changes in institutional practice.

Many anti-rape activists in the 1970s shared not only a faith in law and criminal justice, as Bumiller (2008) and Corrigan (2013a) have argued, but also a hope that technology and science would enhance victims' credibility in the criminal justice system. These hopes were not blind, however. Most recognized that eliminating rape required far more than technological or legal solutions. However, to gain the ear of physicians, scientists, and police, advocates had to speak in institutionalized languages of criminal justice, medical treatment, forensic science, and law. The pressures to conform to government and medicolegal agendas, funding regulations, and institutional requirements redefined and transformed many RCCs into services providers that are more closely aligned with criminal justice processes. Symbolizing this shift, most RCCs now have instructions on their websites on how to preserve forensic evidence, the importance of doing so, the steps of the SAEK exam, and the services that hospital treatment centres provide. In this way, many RCCs' advocacy work now facilitates criminal justice and forensics processes, despite most advocates' understanding that these processes rarely serve victims' interests. Despite the fact that advocates have been replaced in some forensic exam rooms with hospital-based social workers, and RCCs are now rarely involved in the consultations on forensics and medical care for victims, many RCCs have adopted the role of service providers working alongside medical and legal institutions. Under funders' constraints on rape crisis work and enduring funding shortages, many centres have few resources for feminist activism and political organizing. Their capacities to transform cultural systems that maintain and support sexual violence have been hampered by what has been a long and complex relationship with government agencies and medical and legal institutions.

When looking to the future, advocates expressed frustration and hopelessness about medical and legal responses to sexual assault victims. Unlike many of the anti-rape activists in the 1970s who shared optimism that their work in RCCs would end male violence, many contemporary advocates voiced their anger with the lack of change in medical and legal institutions. One described the daily frustrations of her work: "Came in first thing this morning, there is already a call ... I gotta call a woman back who just went through a kit. 'Went to the police and for some reason they are not taking me seriously.' Here we go again. Exact same scenario." Caught under the weight of ongoing demands for their advocacy services and the unchanging practices in hospitals and police stations, many advocates expressed little hope for change.

For some advocates, the kit offers little reason for optimism. Reflecting on their experiences supporting victims in forensic exam rooms and criminal investigations, many voiced strong criticisms of the kit. Despite the optimism with which activists had first envisioned the kit, they said that the contemporary kit turned out to be a "useless tool" that perpetuates an ineffective system that disbelieves victims reporting sexual assault. They described the tool as "absurd," "a joke," and "just another procedure ... that revictimizes [victims'] bodies." One advocate commented on the extent of resources that have been invested into making the SAEK work, from its costs to the resources necessary for SANEs, SACTCs, and forensic labs. She said, "It's sort of hilarious, all this for what?!" Not all advocates, however, were as critical of the kit. A few described it as a "good tool" that has "given some women some comfort" by providing the evidence necessary to move some victims' reports of sexual assault forward in the legal system. They said that it "has made a difference" for some victims. While there were these few comments to the contrary, most advocates suggested that the kit had in fact been turned against victims and their allies, as the advocate at the Jane Doe conference had argued. One advocate explained:

Everything that has been brought forward, everything including the kit, has then turned around and been used against us ... We fought for that and of course ... it doesn't work. And the only reason that none of this works is the institutions themselves have never been ones to understand violence against women, protect women, or believe women. They are set up to maintain the same oppressive structures that create violence against women in the first place.

Reflecting on the relationship anti-rape activists had with medical and legal institutions and the SAEK over the past few decades, another said: "We want to be able to have a voice, to have an impact in shaping it. But how it is used, how it is perceived, we don't have control over that. So you tell yourself that if it is going to have one then you might as well participate to make it the best that it can be, and then whoever is using it will use it according to their own framework." In describing their own history, these advocates revealed how technologies can be reshaped and redefined through practice. They echoed the process that feminist science and technology studies scholar Katie Hasson (2012) and others have described of technologies developing their politics not through their design, but through their use. The kit, in these advocates' eyes, becomes a politicized tool that works against women not because of its material design, but because of how it is used and understood in the criminal justice system.

Despite the many disheartening descriptions of the kit and the medical and legal networks within which it operates, several advocates expressed some optimism for the future. One said, "One thing that will not diminish are feminists. We may have to find a new venue, create some new spaces, but I remain optimistic." Another reflected on anti-rape activism from the 1970s to present and described how future medicolegal practice will be transformed by saying: "There is still a great need for a wave to strike the shore. It's like rain drops, when we strike the shore in tiny tiny bursts you have a momentary impact based on the dents in the sand but I think it does require that wave in order to dislodge some of the very deep-seated thinking and attitudes that are still prevalent."

Counter-networks and a New Wave Striking the Shore

Sexual assault has fallen under mainstream media's spotlight in Canada in recent years, which has sparked heated public debates about institutional responses to sexual assault reports in both the criminal justice system and post-secondary institutions. In 2013, a blaze of news media reports about Canadian universities' responses to sexual violence on campus was ignited when students at Saint Mary's University posted a YouTube video of students at the university singing an orientation chant about sexual assaulting young women and girls. Not long after, Dalhousie University and the University of Ottawa were pushed into the media spotlight when news broke that male students on their campuses had been using Facebook to discuss committing sexually violent acts against their female classmates. News of sexual assaults at other Canadian universities and sexual assault victims being dismissed and ignored by the university administrators to whom they had turned for help made national headlines. Amidst the heightened public attention on campus sexual assault, investigative journalists at major Canadian media outlets turned their attention to the alarmingly high rates of sexual assault on campuses and universities' systemic failure to respond appropriately (Mathieu & Poisson, 2014a, 2014b; Poisson & Mathieu, 2014). Activist groups on university campuses across Canada have since been pushing university administrators to develop sexual assault policies, resource sexual assault services on campus, and address campus rape culture (Quinlan, Quinlan, Fogel, & Taylor, forthcoming).

Fuelling this public dialogue on campus sexual assault was the news of several women coming forward with accusations of sexual assault against two high-profile entertainers: Jian Ghomeshi, a CBC radio host in Canada, and Bill Cosby, a well-known comedian in the United States. As more women came forward accusing Ghomeshi and Cosby of sexual assault, media coverage turned to the systemic nature of sexual violence and the barriers that victims face reporting sexual assault to police. The news inspired the Canadian Twitter hashtag #BeenRapedNeverReported, which generated over seven million tweets in the first few days (Zerbisias, 2014). Under the hashtag, women and men described their experiences of being sexually assaulted and the reasons they never went to police. Fears of not being believed by police and being blamed and re-victimized by the criminal justice process shone through their descriptive tweets. All of this occurred in the aftermath of the international Slutwalk movement, which was sparked in Toronto in 2011 when a police officer told students at York University that women could avoid rape if they stopped dressing like sluts. Beginning with a single protest of over 3000 people in Toronto, the Slutwalk movement gained momentum across Canada and internationally, as people took to the streets to draw attention to the culture of victim blaming and slut shaming that was so clearly epitomized by the Toronto police officer's talk (Teekah, Scholz, Friedman, & O'Reilly, 2015).

More recently, in 2016, Jian Ghomeshi's sexual assault trial and his subsequent acquittal has sparked public outcry about the failures of the criminal justice system to treat sexual assault victims with dignity and respect. Throughout the trial, Ghomeshi's lawyer skirted around existing rape shield laws and attacked the three victims' credibility with methods that critics said mirrored defence lawyer Michael Edelson's tactics from 1988 to "whack the complainant hard" (Schmitz, 1988, 44) (e.g., Doe, 2016; Fraser, 2016; Tanovich & Craig, 2016). The judge's ruling, which acquitted Ghomeshi of all charges and accused the victims of being "deceptive and manipulative" (R. v. Ghomeshi, 2016, 24) and "playing chicken with the justice system" (20), has since generated a wave of anger across the country among victims and their allies. Ghomeshi's trial and its outcome, many have argued, epitomized the enduring legacy of rape myths in sexual assault trials and the continued reliance on legal practices that re-victimize victims of sexual assault (Doe, 2016; LEAF, 2016; Murphy, 2016; Press Progress, 2016).

Amidst public outcry about institutional responses to sexual assault, critical feminist scholars are increasingly calling medical and legal institutions to account for their systemic failure to respond to sexual assault. Mulla's (2014) discerning analysis of the harms of forensic exams, Corrigan's (2013a, 2013b) sharp critique of the failures of sexual assault legal and policy reform, and Sallomi's (2014) critical look at the failures of DNA to prevent sexual violence, are among the vast array of scholarly works calling medical and legal institutions to account. Almost forty years after Clark and Lewis (1977) penned their first critique of sexual assault policing in Toronto and the anti-rape movement spread across Canada and the United States, a new wave of activism and critical scholarship on institutional responses to sexual assault is striking the shore.

The contemporary public dialogues about institutional responses to sexual assault suggest a new consciousness around sexual violence, criminal justice, and sexual assault forensics. Halfon (2010) suggests that struggles against socio-technical networks are most effective when they are supported by "counter-networks" (74). The growing contemporary movement around sexual violence might be understood as a burgeoning counter-network that is being built in response and in opposition to medicolegal networks of experts and forensic technologies for sexual assault. Similar to the kit's medicolegal network, this counter-network is heterogeneous, and includes not only activists, advocates, and scholars, but also the technologies that they are using to increase the volume of critical voices on sexual assault laws, policies, and practices.

As this counter-network draws more attention to institutional responses to sexual violence, greater pressure will fall on medical and legal institutions to reassess their practices. This pressure will likely take different forms than it did in early 1970s, when anti-rape activists were first struggling to transform rape into a recognized social problem. Now, after decades of institutional and legal reform and the resulting professionalization of sexual assault services and depoliticization of many rape crisis centres, the terrain for political action is significantly different. Rape crisis centres may no longer be the hubs of feminist activism that they were in the early 1970s and 1980s. However, if the Twitter hashtag #BeenRapedNeverReported is any indication, social media may be creating new spaces for activists, advocates, victims, and their allies to share stories of injustice and organize for change. It remains to be seen how medicolegal networks might shift under the weight of this new pressure, and how this might influence the SAEK and its current place in medical and legal practice.

Imagined Futures: Reassembling the Technoscientific Witness

Since the SAEK's story began with anti-rape activists imagining better futures for sexual assault victims, it is fitting that they would be involved in any discussion of the kit's possible futures. Rape crisis advocates' unique positioning on the margins of medicolegal practice offers them a perspective on sexual assault treatment, investigations, and prosecutions unlike those of the police, nurses, lawyers, and scientists who are directly involved in this work (Corrigan, 2013a). Advocates' visions of the future provide hints on what new practices, controversies, and assemblages of the kit's network may be on the horizon.

Despite the frustration and disappointment that many advocates expressed in relation to the current SAEK and the network of practices in which it is involved, several expressed hope that new reforms in law and medicolegal practice could change the kit for the better. Some described reforms that they thought had the potential to transform the kit into a more effective and useful tool for victims. Others were less convinced that any reform in practice and policy could change the nature of the kit and its harmful effects on victims. These advocates' contrasting views on whether the kit could be reassembled into a better tool pivoted around the more theoretical question, What defines a technology? Is it defined by the practices in which it is involved, the spaces in which it operates, or the values embedded within it? Feminist scholars have long debated these very questions in the context of discussions about what constitutes a feminist technology (Layne, Vostral, & Boyer, 2010). Advocates that I interviewed revealed very different answers to these theoretical questions as they voiced contrasting perspectives on whether the kit could be reassembled into a more useful tool for victims and how this could be done. The reforms that some advocates described, if realized, would undoubtedly shift the kit's network in significant ways. Assembling new spaces, laws, and tools around the kit would change how it acts and with whom it acts within medicolegal practice. The SAEK's past provides some clues for understanding what effects these future reforms may have on victims, advocacy, and medicolegal practice.

Several advocates suggested that victims' experiences with the SAEK exam might improve if Sexual Assault Care and Treatment Centres were moved out of hospitals and into community rape crisis centres. They suggested that by putting SACTCs in rape crisis centres and community health clinics, victims would have more accessible and less institutionalized care and treatment. This move, they contended, would counter the pressures to professionalize and medicalize sexual assault treatment and would grant the anti-rape movement, victims, and their allies greater control over the SAEK and the practices in SACTCs. For these advocates, a new space held the potential of redefining the kit as a less invasive, and perhaps more empowering, tool for victims. Other advocates, however, adamantly disagreed. One argued that the SAEK will maintain its meaning and function regardless of where it is situated and who works alongside it. She said:

> It is still an evidence kit for the legal system. So you would still have to provide what the legal system wants you to provide ... What I would see happening there, is the sexual assault centres would then be used as a guise to look like they were helping women, but in actual fact they are just replicating what the system is already doing to women, with a guise of it being a nicer place ... That would be forcing sexual assault centres to be an institutional space that we don't want it to be.

For this advocate, and others who expressed similar sentiments, a new space would not redefine the kit. The shaping effect would go the other way; instead of the space redefining the kit, the kit would redefine the space. The kit would transform rape crisis centres into forensic spaces that reflected the values and priorities of the legal system. Within advocates' contrasting views about a new space for the SAEK are very different notions about what defines the contemporary kit.

While these advocates spoke about kit exams in Canadian rape crisis centres in the abstract, their vision has become a concrete reality for some community-based rape crisis centres in the United States. The vast majority of sexual assault forensic exams in the United States are conducted in hospitals; however, some rape crisis centres have hired Sexual Assault Nurse Examiners to offer forensic exams and medical care to victims. Corrigan (2013a, 2013b) argues that because of their apparent distance from hospitals and law enforcement agencies, these community-based programs are often more effective in supporting victims, collecting high-quality evidence, and protecting victims' legal rights. By aligning themselves with SANE programs, these RCCs also seemingly have greater political leverage and power to influence and challenge practices in law enforcement. However, Corrigan also notes that these changes can come at the cost of obscuring feminist analyses of sexual assault that many RCCs were built upon and reinforcing medicalized responses to sexual assault. With these changes happening south of the Canadian border, Corrigan's findings may point to possible changes that could be on the horizon for some Canadian rape crisis centres. The kit's history provides some insight for understanding the possible consequences of these changes.

As I charted in chapter 3, when SAEK exams moved from Ontario emergency wards into SACTCs in the 1980s and 1990s, new actors, expertise, expert spaces, and material relations were created. The kit and its new space were co-constituted: the new hospital space redefined the kit as a highly specialized tool for expert use, while the kit defined the hospital space as one that served the purposes of law and forensic science. The shaping effect thus went in both directions. This history illustrates how changing spaces for technoscientific objects can destabilize and reassemble relations in which the objects are involved. Moving the SAEK out of hospitals and into the RCCs may redefine (at least in part) sexual assault treatment and care as a less medicalized practice. Likewise, the kit may be redefined as a tool that advocates have more control over, at least during the forensic exam. However, if the kit is indeed an actor with the capacity to act and affect actors around it, as I have argued in this book, it would also shape relations within RCCs. Rape crisis centres would have to redefine themselves as spaces that could facilitate and support legal requirements for SAEK evidence. For those victims, advocates, and activists who rely on RCCs as alternatives to medicolegal spaces, this shift could have negative consequences. Victims would have fewer collective spaces untouched by the SAEK and its forensic script and RCCs would be burdened with the new pressures to fund SAEK exams (in part or in full) and build the credibility and appearances of objectivity that lawyers and judges require for SAEK evidence. When SACTCs were developed, proponents argued that the new space for the kit would vastly improve victims' experiences in the exam. While this may be true for some victims, reports of victims' negative experiences with the SAEK exam suggest that these centres have not lived up to their promise for all victims. In a similar way, moving the kit into the Canadian RCCs may not have empowering effects for all victims. Instead, the results will likely be mixed and will require critical analyses of whose purposes they actually serve.

Other reforms that advocates imagined focused on law and legal practice. One advocate envisioned a legal reform that would change how the SAEK is used in sexual assault courtrooms. She described a *kit shield law*. This law, she suggested, would run parallel to the rape shield law for sexual history developed in the 1990s and would prevent defence lawyers from using a victim's choice to have the SAEK exam (or not) as evidence in sexual assault cases. Changing some of the legal practices around the kit, in this advocate's view, would redefine the ways that the kit can act in the courtroom and limit some of its negative consequences on victims. After recounting stories of anti-rape activists organizing in the 1990s to implement the rape shield law for sexual history, she said she hoped to begin organizing contemporary activists to press for a kit shield law.

A kit shield law would restrict the ways that the SAEK could be read in the sexual assault courtroom. It could prevent defence lawyers from publicly drawing inferences about a victim's truthfulness from her choices about the SAEK. In so doing, it may reduce some of the pressures that police, nurses, and others put on victims to consent to SAEK exams. While this legal reform may curb some of the negative ways that the kit acts on victims in the courtroom, it may have less of an effect on practices outside the courtroom. The rape shield law enacted the 1990s had some success in limiting the evidentiary use of sexual history in sexual assault cases (Johnson & Dawson, 2011), although, as Ghomeshi's recent trial reveals, defence lawyers have found ways to work around it (Brean, 2016). Regardless of its mixed success in court, as revealed in chapter 4, the rape shield law has done little to restrict police investigation of victims' sexual history and the inferences drawn from it. In a similar way, a kit shield law could be expected to do little to restrict police practices of reading a victim's consent to the SAEK exam as an indication of her truthfulness.

Other advocates imagined more radical changes to the SAEK itself. One such imagining emerged as a survivor was describing the contemporary kit. She said: "I think it [the kit] is a damned if you do, damned if you don't situation for women. But if there is a way we could twist that dynamic. Should the focus be on women? Or should the focus be on the accused? [*pause*] Could we do a forensic exam of him? [*laughs*]" For this survivor, a sexual assault kit for an accused was an ironic twist on a technology that, more often than not, causes victims pain and trauma. By reimagining the SAEK, she envisioned a new set of medicolegal practices that would shift the discomfort and pain that the kit can cause away from victims and onto those accused of sexual assault. For her, the absurdity of this reimagined kit was laughable. However, a SAEK for the accused is perhaps not that far out of the realm of possibilities.

Although there is limited forensic scientific literature on the topic, some researchers have begun to examine rates of retrieving female DNA from male bodies following sexual activity. One team of researchers recovered female DNA in 100% of penal swabs taken within twenty-four hours after intercourse (Farmen, Haukeli, Ruoff, & Frøyland, 2012). Another team recovered female DNA from three-quarters of the male fingernails that were scraped within twelve hours after digital penetration (Flanagan & McAlister, 2011).[3] Just as scientific findings contributed to fuelling the development of the SAEK, these emerging scientific findings might do so for evidence kits for the accused. Collecting forensic evidence from suspects in sexual assault cases is a practice well established in some areas of the United States. In Canada, this development seems less likely under current search and seizure provisions in Canadian law; however, there are hints of an emerging turn.[4]

One police officer I interviewed described a recent case where she and her co-investigator took a penile swab without a warrant from a suspect of sexual assault. Despite successfully obtaining DNA evidence from the penile swab, the police officer expressed some hesitations about how it might be received in court.

> This has not gone to court yet so maybe I'm going to get grilled on the stand, maybe they won't admit it into evidence. They'll say, "Oh no you need a warrant" … As long as you are not …, what the heck is the terminology they use? It can't be an intrusion of privacy … We are going to

argue that this is not an invasion of privacy, this is not an invasive type of procedure, there was no pain caused to him, it was very simply a swipe with a Q-tip more or less. His lawyer will argue that it shouldn't be admitted. But, we'll see how that pans out.

Cases like these could inspire the creation of new Canadian forensic kits for accused. As the history of the kit and DNA evidence suggests, these kits would likely stir up much legal debate and controversy.

The current SAEK has done little to decrease rates of sexual assault and influence the outcomes of sexual assault cases (Johnson, Peterson, Sommers, & Baskin, 2012; McGregor, Du Mont, & Myhr, 2002). In a similar way, a sexual assault kit for the accused may in fact have a negligible impact on rates of sexual assault and sexual assault case outcomes. Several advocates described how perpetrators of sexual assault are becoming more aware of the possibilities of being identified through forensic evidence-collection practices. This awareness, they suggest, has not deterred offenders from committing sexual assault. Instead, it has inspired perpetrators to sexually assault victims in ways that are less likely to leave forensic traces, such as using condoms so that their semen is less likely to be detected and collected. A sexual assault kit for accused may act in similar ways. While this new kit could shift forensic practices and redefine the nature of the forensic exam, it may also replicate and reproduce many of the limitations of the SAEK.

Regardless of their impact on rates of sexual assault, all the reforms and changes that advocates described have the potential to disrupt, destabilize, and potentially reassemble the SAEK and its medicolegal network. Small changes in practice and material objects will likely have ripple effects into other practices and relations between human and non-human actors involved in sexual assault response. The extent to which these ripple effects will benefit all victims, however, is less clear.

Johnson (2010) argues that technologies can be considered *feminist* if they work within social relations that reflect gender-equitable arrangements. For a technology to be feminist, she argues, the material design of the technology, as well as the social and technical systems of which the technology is a part, must be oriented to enhancing gender equality. By this definition, the history of the SAEK reveals that the kit is not, nor has it ever been, a feminist technology. Although the kit was in part inspired by feminist advocacy and feminist critiques of medical and legal practice, it does not reflect these origins. Emerging out of patriarchal and racist legal histories of doubting women's reports of rape, the SAEK continues to perpetuate and reflect a system in which rape victims have little credibility and technoscience is used to exclude, coerce, and interrogate them. Reassembling the technoscientific witness will require more than just reassembling the kit's material design, because that design is a reflection of the inequitable and ineffective system in which it operates. The kit's history is a reminder of this and a testament to the challenges of reassembling technoscience and imagining better futures.

"It could have been otherwise"

The SAEK's history is often forgotten. When policymakers, legislators, police, and lawyers sing the kit's praises for improving legal responses to sexual assault, its humble historical origins in the pages of feminist literature and at the consultation tables of sexual-assault task forces are rarely recounted. But the kit's past is significant to its present. Bound up in the kit are entangled histories of law, medicine, science, technology, and feminist activism that are crucial to understanding how and why police, nurses, and lawyers respond to victims of sexual assault in the ways that they do. Tied to the kit's history are histories of sexist and racist rape myths that have pervaded the law and legal practice in sexual assault cases for decades and the histories of feminist activism that sought to resist them. By diffracting the technoscientific witness of rape, it becomes possible to see the kit as a tool that is not innocent. Instead, it is a tool that retains the marks of a legal and medical system that has a long history of dismissing, doubting, and blaming victims of sexual assault. It is this historical vision of the kit that is necessary for informing and inspiring future demands for change.

The SAEK's history reveals that technological, scientific, and legal solutions to sexual assault have done little to transform the systems that maintain sexual violence and the institutions that respond to it. Despite the elaborate network of scientific, medical, and legal practices dedicated to responding to sexual assault, little has changed. Advocates are still plagued with the same questions they were forty years ago about how institutions can be made more accountable, accessible, and responsible to victims. And many victims are still reluctant to turn to these institutions that claim to work in their interests. Corrigan's (2013a) argument that this is largely a result of the anti-rape movement placing too much faith in law and criminal justice to eradicate sexual violence can be extended even further. Too much faith has been stored in not only law, but also science and technology to solve the problem of sexual violence and the failures of medical and legal institutions to adequately and sensitively respond. By looking into the kit's past, it becomes possible to see that future progressive change will require moving away from technoscientific and legal solutions to sexual violence and, instead, reassembling the practices, politics, and values of the institutions that respond to sexual assault.

The SAEK "could have been otherwise" (Hughes, 1971, 552), if controversies had ended differently and practices, technologies, and networks had been assembled in different ways. Seeing the kit's origins reveals that its shape and meaning were never inevitable; instead, they were the products of controversy, negotiation, and struggle. The kit is an actor that was influenced and shaped by its relations with other actors. If the kit *could* have been otherwise in the past, then it follows that it *can* be otherwise in the future. Haraway (1994) suggests that the task of feminist technoscience studies is to examine how technoscience can be otherwise, to facilitate the imagining of more just, ethical, and responsible ways of assembling technoscience. Locating the histories of the SAEK's design and use in particular actors and networks demands a renewed accountability for the kit and the way in which it acts. It also opens up new possibilities for imagining how the kit could be made to act in different ways.

Although I have argued that the kit has agency in medicolegal networks, I have not done so to diminish the responsibility of the human actors who act with it. Haraway (1997) reminds us that although nonhumans can act, it is human actors who have the "emotional, ethical, political, and cognitive responsibility inside these worlds" (10). Human actors have the responsibility to envision and enact alternative and more ethical modes of technoscientific assembly. For the SAEK to become a tool that does not rest on the private suffering of victims, human actors will have to reassemble the relations that make it what it is. It is a task that requires the recognition of the kit's troubled past and optimism for a better future.

Appendix

Interview sample

Participant type	Number
Advocates in rape crisis centres	14
Sexual assault treatment centre staff	9
Forensic scientists	4
Police investigators	13
Police administrators	4
Police victim services	5
Criminal lawyers	13
TOTAL	62

Participants' years of experience in field

Participant type	Average years of experience in field
Advocates in rape crisis centres	18
Sexual assault treatment centre staff	9
Forensic scientists	29
Police investigators	12
Police administrators	18
Police victim services	5
Criminal lawyers	20

Notes

1 Introduction

1 My use of the dated term *rape crisis centre* is deliberate. Throughout this book, I use the term rape crisis centres to refer to community-based organizations that provide advocacy, public education, and support for victims. Since legal reforms in 1983 that removed rape from the Criminal Code of Canada, many centres have adopted new names, such as sexual assault crisis centres and sexual assault survivors' centres. For clarity, however, I use the name rape crisis centre. I use *rape crisis advocates* to refer to staff and volunteers who support those who have experienced sexual assault. I also use the broader term, *anti-rape activists*, to refer to feminist and women liberation organizers, scholars, lawyers, and advocates involved in political work around sexual violence.

2 Most other jurisdictions in Canada use the sexual assault kit designed and distributed by the Royal Canadian Mounted Police.

3 For some well-known examples, see Aronson (2007), Cole (2001), Gerlach (2004), Jasanoff (1995), Latour (2010), and Lynch, Cole, McNally, & Jordan (2008).

4 See Epstein (1996), Hoffman (1989), and Kling & Iacono (1988).

5 The concept of boundary objects is clearly articulated in Star and Griesemer's (1989) study on the Berkeley Museum of Vertebrate Zoology. This study focuses on the heterogeneity of scientific work and outlines the various boundary objects that amateur collectors, professionals, and museum administrators used to facilitate their cooperation and communication. By drawing attention to boundaries as they relate to scientific practice, Star and Griesemer pick up a similar theme to Thomas Gieryn (1983), who a few years earlier described "boundary work" (781)

as scientists' efforts to draw and maintain boundaries between science and non-science. While similar in theme, there are some important differences between these concepts. While Gieryn uses boundary work to discuss the maintenance of boundaries *around* science, Star and Griesemer use boundary objects to describe the heterogeneity *within* science and the objects that are used to transcend it. For more discussion on the distinctions between these two concepts see Riesch (2010), and for more specific discussion on boundary objects, see Star (1989, 2010).

6 White has also published under the name Parnis.

7 In forensic contexts, anthropometry involves the systematic measurement and analysis of parts of the human body, such as height and shoe size, for the purposes of identification.

8 For more discussion on actor-network theory and its usability for studying crime and crime control, see Robert & Dufresne (2015b).

9 Sexual assault nursing and sexual assault policing are similarly affected by high turnover rates. Both fields involve on-call work, close contact with trauma, and pay that for many is marginal compensation for the emotional challenges and exhaustion of the work.

10 For more discussion on the challenges of gaining access to medicolegal spaces, see Parnis, Du Mont, & Gombay (2005).

2 Inscriptions of Doubt

1 Throughout this chapter, I use the term *rape* instead of sexual assault to reflect the historical context in which the kit emerged. In the 1970s and early 1980s, rape was an offence in the Criminal Code of Canada and was narrowly defined as forced vaginal penetration. During this time, feminist activists and scholars commonly used the term rape in public education and feminist literature (e.g., Brownmiller, 1975; Medea & Thompson, 1975; Clark & Lewis, 1977). Accordingly, many of the public discussions around medical and legal responses to sexual violence focused on rape and not other forms of sexual violence. This is the context that my language usage reflects.

2 Although the women's movement of the 1960s and 1970s had opened the doors to more women in law, medicine, and science in the 1970s and 1980s, these three professionalized fields remained largely dominated by white men (Brockman, 2001; Feldberg, Ladd-Taylor, Li, & McPherson, 2003).

3 In 1975, the Criminal Code of Canada was amended to prohibit this line of questioning without the permission of the judge. However, historical evidence suggests that this practice continued (Clark & Lewis, 1977).

4 The rape conviction rates for rape in the 1970s were drastically lower than those for other crimes. According to one report, the *acquittal* rate for most crimes was 15% in the mid-1970s, whereas the *conviction* rate for rape cases was 15% (CBC, 1971). Clark and Lewis's (1977) estimate of the likelihood of being convicted of rape in Toronto was even lower, at 7%. They argue that low conviction rates in rape cases were a function of both practices in the courtroom and in police investigations that hinged on beliefs that women lied about rape.

5 The belief that rape resulted in physical injuries had its origins in 18th- and 19th-century English law, where rape was defined as sexual intercourse with a woman that was "by force and against her will" (MacFarlane, 1993, 16), a view that was propagated by medical expert claims that it was impossible to rape a woman without extreme force, such as the claim in the *Elements of Medical Jurisprudence* in 1815, which stated "you cannot thread a moving needle" (as cited ibid., 20). MacFarlane's historical research reveals that medical evidence of vaginal penetration and injurious marks from force and violence was commonly required for conviction in rape cases.

6 Some of these techniques included (a) the scanning electronic microscope, which unlike its predecessor, the optical microscope, could magnify objects up to 300,000 times, (b) gas chromatography, a more precise technique for detecting and separating chemical compounds, and (c) mass spectrometry, a more sensitive technique for identifying chemical compounds (Krishnan, 1978).

7 While the 1960s and 1970s were decades when feminist consciousness around male violence solidified, the movement had historical roots that were much deeper. Gavey (2005) estimates that women's organizing against male violence and sexual exploitation began as early as the 19th century. Much of women's early activism around sexual violence in Canada is not well documented; however, Backhouse (2008) notes that in the 1900s, some women were challenging existing rape laws and organizing for legal reform.

8 As I discuss in the following chapter, radical feminist analyses of rape were heavily criticized in later years for ignoring the differences in women's experiences of rape and the ways that sexual violence is motivated by not only patriarchy, but other systems of oppression, such as racism, classism, homophobia, and ableism (e.g., Davis, 1983; Harris, 1990; Monture-Okanee, 1992).

9 The OCRCC requested yearly funding of $15,000 for each of the 15 centres for three years (Ontario Coalition of Rape Crisis Centres, 1979). The PSJ, the predecessor to the current Ontario Ministry of Community Safety and

Correctional Services, declined the OCRCC's first funding proposal in February 1979 with a recommendation that the coalition seek funding from community agencies instead (Rape Crisis Centre may have to close, 1979). The coalition publicized the PSJ's decision in the media and applied again in November 1979. In response to public pressure, the PSJ conceded to negotiate.

10 Apart from general descriptions of the 1978 SAEK design consultations (CARSA, 1980; No author, ca. 1981), I found no historical records on the specific discussions that occurred within the consultations. If meeting minutes were taken at these consultations, they were not included in the archival records on the SAEK and the anti-rape movement that are now scattered across collections in the Archives of Ontario, University of Ottawa Archive and Special Collections, Miss Margaret Robins Archives of Women's College Hospital, and City of Toronto Archives. Despite the absence of these historical records, existing records on the relations between RCC advocates and institutional actors discussed in the previous sections of this chapter offer some insight into what may have occurred at the design consultation table. In this section, I draw on my broader analysis of the relations between the anti-rape movement and medical and legal institutions to paint a picture of what may have occurred during the consultations on the SAEK's design.

11 Not all aspects of the victim's report of rape were excluded from the kit's contents. While the victim's subjective emotional and psychological experience of rape was largely excluded, the kit's standardized forms included physicians' descriptions of the victim's visible emotions and the specifics of the rape. It is these aspects of the rape that were translated then into forensic language. Research on the contemporary forensic exams has shown that victims' reports of rape often guide how forensic examiners conduct the exam (Parnis & Du Mont, 2002, 2006; Mulla, 2014). This evidence suggests that victims' reports of rape can play a role in the context of contemporary exams. However, as I discuss in the next chapter, the early SAEK did not allow for this level of discretion among examiners. The early kit instructed physicians and nurses to follow the kit's standardized steps regardless of the specifics of the rape, the victim's emotional state, or the extent to which the staff believed the victim (Procedures form, ca. 1981). The victim's narrative thus likely played a less significant role in the context of the early forensic exam than it does in contemporary exams.

12 This changed several years later, when, according to a forensic science administrator I interviewed, the PSJ switched to a private contractor for SAEK preparation.

13 The standardized SAEK had been in use since 1979 (Provincial Secretariat for Justice, 1979a), but was formally introduced in 1981. Why the announcement of the SAEK was delayed by two years is not clear in the historical records.

3 Stabilizing the SAEK

1 Other reports expressed similar sentiments, claiming that the SAEK would "catch more rapists" (Kit will catch more rapists, 1981, A24), "help prosecute offenders [and] avert dismissals" (In brief, 1981, P5), and ensure "victims' rights" (Speirs, 1981, CL8).

2 I use the terms rape and sexual assault interchangeably in this chapter to reflect the changing language in the 1980s. Rape was eliminated from the Criminal Code of Canada in 1983 and replaced with three tiers of sexual assault offences. I use the term sexual assault to reflect this reform and the term rape to illustrate the political potency that the term rape maintained for many activists in the anti-rape movement.

3 The funding conditions and requirements that the PSJ imposed were similar to those that other government funding agencies placed on other women's organizations and services (for more discussion, see Cohen, 1993; Ng, 1996).

4 Two centres that have retained their collective structure are Toronto Rape Crisis Centre/Multicultural Women Against Rape and the Sexual Assault Support Centre of Ottawa.

5 For an in-depth discussion of similar trends in RCC history in the United States, see Corrigan (2013a) and Bumiller (2008).

6 This term includes both the anti-rape movement and the movement centred on battered women (later termed domestic violence).

7 Omitting SAEK steps was permitted; however, this could only be done if an investigating officer and physician agreed it was necessary and a detailed rationale for the omission was provided in the SAEK documents (Taking care, 1990).

8 For many victims, consenting to the SAEK exam could be complicated by many things, such as police pressures and lack of information about the SAEK (see chapter 6 for further discussion). From the original SAEK consent form, it is unclear whether victims had the right to consent to a SAEK exam without immediately reporting a rape to the police. Although the consent form suggests that the decision to report and to have the exam were separate, the form provided no way for her to specify to which of the two options she was consenting (Consent form, ca. 1981). The consent

section on the Sexual History Form (ca. 1981) created further confusion by combining the consent to the SAEK exam with consent for medical personnel to report the rape to the police. Consent to the SAEK exam was thus tied to consent to police involvement.

9   Understanding the sexually violated body as a crime scene in forensic examination is a common dictum reiterated in nursing and psychology literature (Price et al., 2010; Johnson, Peterson, Sommers, & Baskin, 2012; Campbell, Patterson, & Bybee, 2012). Feminist scholars have explored how this understanding disempowers women who experience rape (Doe, 2012), reconfigures sexually violated bodies and the domestic spaces they occupy (Mulla, 2008, 2014), and fragments the body and reduces conceptions of rape to bodily violence (Bumiller, 2008).

10   While there has been no historical analysis of whether the SAEK did indeed increase the admissibility of medical forensic evidence in rape cases (Du Mont & Parnis, 2003), one Crown prosecutor who practised in the 1970s and 1980s suggested that after the SAEK was developed, medical forensic evidence was rarely challenged in court and was commonly deemed admissible.

11   Susan Estrich (1986) describes "real rape victims" (1088) as victims whose identity and rape fits the sexist stereotypes that define the rapes police choose to believe.

12   This would all significantly change with the introduction of DNA analysis in the late 1980s, which I discuss further in the next chapter.

13   For examples, see Training manual (1977), Toronto Rape Crisis Centre (1979), Sudbury Regional Rape Crisis Centre (ca. 1980), Sexual Assault Centre London (ca. 1981), Sexual Assault Crisis Centre Windsor (ca. 1984).

14   These included: (1) sexual assault, with a maximum penalty of 10 years (s. 246.1), (2) sexual assault with the use or threat of bodily harm (s. 246. 2), with a maximum penalty of 14 years, and (3) aggravated sexual assault, with a maximum penalty of life imprisonment (Bill C-127, 1982).

15   Physicians' scope of practice gave them the authority to collect medical samples from body orifices, which was required for the SAEK exam (Macdonald, Wyman, & Addison, 1995, 1). Additionally, physicians were chosen to conduct the SAEK exam because it was assumed that victims would be more likely to disclose rape to a physician and that physicians were more likely to be seen as credible witnesses in rape cases (Martin et al., 1985).

16   Despite not having the same degree of authority in the forensic exam as physicians, from the historical reports, it appears as though nurses received more training on the SAEK than physicians: in the Ontario

Hospital Association survey in 1983, only 65% of Ontario hospitals had trained physicians on the SAEK, whereas 78% had trained nurses (A report of the OHA survey, 1983).

17 The idea for a specialized sexual assault treatment centre was in existence long before it was suggested in Toronto. Hospital-based sexual assault treatment centres had been running in the United States since 1972 (Burgess & Holmstrom, 1973). In Ontario, a regional rape treatment centre had opened in Hamilton in 1979. When the steps were undertaken to design and build a regional sexual assault treatment centre in Toronto, these existing centres were used as models.

18 Under the SACTCs' expanded mandate, many were renamed as Sexual Assault/Domestic Violence Treatment Centres. For simplicity, I use the name Sexual Assault Care and Treatment Centre (SACTC).

19 Sexual assault treatment centres in Ontario introduced French services in 1994 (Ontario Women's Directorate, 1994). I found no evidence of other language services being offered before this date.

20 In the 1980s, lack of accommodation for individuals with physical impairments was not unique to SACTCs and characterized many institutional and public spaces (Ringaert, 2003). In SACTCs, accessible examining tables that accommodated mobility impairments caused by disability and severe injury were eventually introduced in 1998 (Barnhouse, 1998, D. Barnhouse to P. Campbell, 2 March 1998).

21 I draw the contents of this description of the SACC from the SAEK training video *Taking Care* (1990), which was filmed at the Women's College Hospital SACC.

22 This change coincided with many other expansions in nurses' scope of practice (Worster, Sardo, Thrasher, Fernandes, & Chemeris, 2005).

23 *R. v. Mohan* sets out four criteria for expert evidence, of which this is one. The other three criteria are: the evidence must be "relevant"; "necessary to assist the trier of fact"; and "there must be no exclusionary rule otherwise prohibiting the receipt of the evidence."

24 Not all hospitals in Ontario have SANE programs, and thus there are hospitals where physicians and nurses without SANE designations use the SAEK. I describe this further in chapter 6.

4 Assembling the Genetic Technoscientific Witness

1 Forensic DNA typing is not limited to sexual assault investigations. It was and continues to be used in a variety of criminal investigations such as murder, manslaughter, and robbery, where there are traces of bodily tissues

or fluids. This chapter focuses on the history of DNA typing as it relates to sexual assault investigation.

2 Here, I am not suggesting that action in other locations ceased. Instead, I am proposing that during the rise of DNA typing, there was heightened activity and controversy in these locations around the SAEK and sexual assault.

3 The term "second assault" emerged in sexual assault literature in the 1980s and was used in the 1990s to describe the traumatizing and re-victimizing effects that the criminal justice process has on many victims (Campbell & Raja, 1999; Madigan & Gamble, 1991; Martin & Powell, 1994; Williams & Holmes, 1981). The violent metaphors that Edelson employs to describe his defence strategy for challenging a victim's credibility provide a vivid illustration of how courtroom processes could become a veritable second assault.

4 This is perhaps somewhat overstated. As I described in chapter 3, in 1976, anti-rape activists drafted recommendations for rape law reform for the Law Reform Commission (Vance, 1978). However, the extent to which the government solicited activists' recommendations in 1976 is unclear from the historical records.

5 Another significant legal development in the 1990s was what has come to be known as the "no means no legislation," where the Supreme Court ruled that implied consent is not a valid defence for sexual assault (R. v. Ewanchuk, 1999).

6 Here I am drawing on the notion of translation often used in actor-network theory as a moment of transfer where new relations are created in a network (Law, 2006).

7 Chemical analysis was also used; however, most forensic scientists considered it to be complex and unfruitful because the chemical composition of hairs varied across the body (Krishnan, 1978).

8 DNA analysis was also sparked by growing discontent with other forensic methods, such as forensic fingerprinting. For a detailed history of the controversies surrounding forensic fingerprinting, see Cole (2001).

9 This term was later rejected by forensic scientists and geneticists because it was thought to obscure a crucial difference between forensic fingerprinting and DNA profiling: a DNA profile's uniqueness is determined through statistical probabilities of a DNA profile's rarity within a population, whereas a fingerprint's uniqueness is assumed (Cole, 2001). While fingerprints were thought to be a unique mark of individuality, DNA profiles are seen as rare genetic sequences (ibid.). The terms "DNA profiling," "DNA typing," and "DNA analysis" were adopted as

alternatives to Jeffreys's original "DNA fingerprinting." In accordance with this trend, I use these three terms interchangeably.

10 The scientists rested the technique's revolutionary potential in rape cases on a narrow understanding of rape as an act of penile-vaginal penetration perpetrated by an unknown offender.

11 A complete RFLP DNA analysis required approximately 10–12 weeks and approximately 10 ml of sperm or a bloodstain the size of a quarter (Campbell, 1996).

12 For a detailed anthropological study of PCR's development, see Paul Rabinow's (1996) *Making PCR*.

13 Many of these courtroom debates occurred during admissibility hearings, where the scientific validity and relevance of DNA evidence was debated. In the United States, expert evidence was evaluated under the Frye "general acceptance standard," which stipulated that expert evidence used in court had to be relevant and generally accepted in scientific communities. The extent to which DNA evidence fit the Frye criteria was the subject of much legal controversy (Lynch et al., 2008).

14 By this point, a threshold of admissibility for novel scientific evidence had been established through R. v. Mohan (1994), which was used to determine the admissibility of DNA evidence in subsequent cases.

15 In the early 1990s, DNA samples could only be obtained directly from suspects with their consent. In 1995, Bill C-104 was passed, which permitted police to obtain a warrant to seize biological samples from suspects for forensic testing. Under Bill C-104, it became possible to compare a DNA sample from a SAEK with a DNA sample from a suspect that was forcibly obtained with a legal warrant. Attempts to challenge the bill as a violation of suspects' rights under the Canadian Charter of Rights and Freedoms were unsuccessful, and it was praised as advancing justice and public safety in Canada (R. v. S. F., 1997).

16 As of February 2017, the databank included 342,436 convicted offender profiles and 127,787 crime scene DNA profiles (Royal Canadian Mounted Police, 2017). These statistics are continually increasing, as the databank receives an estimated 500 to 600 samples per week.

17 There are few publicly available records on this working group, who was part of it, how the group functioned, for how long, and to what end. The records I have found, which are cited in this section, include only short summaries of what was decided and do not include the list of attendees. Around the same time the working group formed, the provincial coordinator of the sexual-assault treatment centres became the primary contact for the CFS forensic scientists who were redesigning the SAEK.

This marked a significant change from the SAEK's early history, when rape crisis advocates were included in SAEK design meetings. From this, it could be assumed that SANEs took a central role in the SAEK Working Group.

18 CFS, "Sexual Assaults and Forensic Evidence: SAEK Consultation" (29 May 2001), obtained through FOI request to the Centre of Forensic Sciences, no. CSCS-A-2012–01825.

19 Ibid.

20 CFS, "Hospital Instructions" (2012), obtained through FOI request to the Centre of Forensic Sciences, no. CSCS-A-2012–03930.

21 CFS, "Sexual Assaults and Forensic Evidence: SAEK Consultation" (29 May 2001), obtained through FOI request to the Centre of Forensic Sciences, no. CSCS-A-2012–01825.

22 Some SAEK evidence that did not dry easily was still frozen, such as sanitary napkins, condoms, and wet diapers.

23 Ibid.

24 The systemic disbelief of women's experiences of sexual violence in contemporary policing is pervasive and well documented (Crew, 2012; Doe, 2003, 2012; Dellinger Page, 2010; Flood & Pease, 2009; Irving, 2008; Johnson, 2012). Ten out of the twelve sexual-assault police investigators that I spoke with discussed false accusations of sexual assault as a relatively common occurrence. I discuss this more in chapter 6.

25 The National DNA Databank claims to have assisted with more than 11,000 investigations, 1500 of which involve sexual assault (Standing Committee on Public Safety & National Security, 2009).

26 Turnaround time is measured by the time required to conduct the analysis and issue a final report. It is important to note that this number does not reflect the total range of turnaround times. In 2011, only 26% of the sexual assault cases analysed at CFS were completed within the 60-day target date (Centre of Forensic Sciences, 2011). The length of delay for the other 74% of cases was not reported.

27 These numbers pale in comparison to those in the United Kingdom, where there is a two-to-ten-day turnaround time for DNA typing at the national lab that operates 24 hours a day, seven days a week and employs 2500 staff (Auditor General of Ontario, 2007).

28 Quebec and Ontario are the only provinces in Canada with their own provincial forensic lab. All other provinces use the RCMP Forensic Laboratory Services.

29 Given the provincial government's pressures on CFS to explore "cost saving opportunities" (Standing Committee on Public Accounts, 2008) and

the more recent cuts and layoffs in the CFS's electronics unit (Bonokoski, 2012), a budget increase appears unlikely.

30 The number of jobs that were sacrificed to achieve this number was not reported.

5 Instability Within

1 By beginning with the question *who benefits from* the SAEK, I follow in Jane Doe's (2012) path, who, in a recent book chapter, began her investigation of the SAEK with the same question. Doe's work explores this question through victims' and advocates' perspectives on the kit. This chapter takes a different approach. Here, I explore who benefits by interrogating medicolegal practice: who does what, when, where, how, and why?

2 "TV behaviour," according to Baeza and Turvey (2002), includes anything that could be defined as "mimicking the way that stereotypical victims act on television ... (hysterical, demanding female officer, catatonic, etc.)" (177).

3 These criteria have been paraphrased and assembled from multiple interviews with police investigators. For a more detailed discussion of this data see Quinlan (2016).

4 Some interviewed nurses stated that they have an explicit policy not to allow rape crisis centre workers, or any other support people, to be in the forensic exam room. Others described how social workers in SACTCs have now taken on the role of advocate in the exam. These policies and practices are one of ways that rape crisis centre workers have been displaced in practices involving the SAEK.

5 Some centres rely on physicians to order blood work and prescribe medications, while others operate on the basis of medical directives that give SANEs/RNs the authority to provide those options.

6 Many participants in this study, including several advocates, used this term during their interviews.

7 This number reflects the average number of patients the Ottawa SACTC sees per month (Louisa, 2010, A1).

8 For example, during 2008–9, the number of full-time equivalent positions at the SACTC at Trillium Health Partners hospital dropped from 8.13 to 6.47, as did their medical services budget, from $310,912 to $193,887 (a drop of $118,025). What was lost through these cuts was partially regained by 2012–13, when the number of full-time equivalents was raised to 7.3. (THP, "SADV Budget" (2012), obtained through FOI request to the Trillium Health Partners, no. 13-008).

9 In 2011, the Ontario government released a Sexual Violence Action Plan, which promised $15 million over the next four years for service provider training, public education, and prevention (Ontario Women's Directorate, 2011). Although the plan promises increased levels of training for SACTC medical staff and counsellors, it does not include any allocated funds for this purpose. In fact, none of the $15 million was explicitly allocated to SACTCs.

10 In Ontario, most RPNs have a two-year college diploma, whereas most RNs have a four-year undergraduate degree.

11 RPNs' scope varies between centres. However, one centre has RPNs conducting all parts of the SAEK exam, including the pelvic exam. Prior to 1995, RNs did not have the authority to do so.

12 Feldberg (1997) and McGregor, Du Mont, and Myhr's (2002) studies are the only two listed here that were conducted in Ontario. Others listed were conducted in the United States. These regional differences may be significant when interpreting these studies' results.

13 I use this term here, instead of victim, to reflect how these participants self-identify.

14 Parnis and Du Mont (2006) have shown that forensic nurses and police have varying interpretations of the SAEK and routinely assign different meanings to the technology and its evidence. They describe how forensic nurses interpret the SAEK's instructions differently and how some police attribute more value to the SAEK than others. By thinking about the SAEK as an actor that becomes multiple in practice, my analytic gaze shifts beyond human actors assigning meaning to the SAEK towards the practices in which the SAEK is involved.

15 By using the phrase "point of action," I am not implying that medicolegal practice is singular. Rather, I understand these points of action to include an assembly of actors and actions around a particular site in the SAEK's travels through the medicolegal network.

16 In sketching the access restrictions on the SAEK, I am not arguing that it is necessarily a tool that *should* be accessed. Rather, I draw out these practices to investigate how and for whom the kit is enacted.

17 There are some exceptions to this, particularly in some remote Northern Ontario communities, where access to urban SACTCs is severely restricted. In some of these communities, SACTC nurses are providing outreach training for nurses who conduct the SAEK exam.

18 For more discussion on the effects of institutionalized racism on Indigenous peoples' access to health care in Canada see Allan & Smylie (2015) and Tang & Browne (2008).

19 See chapter 3 for further elaboration.

20 For a more detailed discussion of how SAEK evidence can influence sexual assault trials, see Du Mont & White (2007), Johnson, Peterson, Sommers, & Baskin (2012), and McGregor, Du Mont, & Myhr (2002).

21 In the 2001 revision of the SAEK, two separate consent forms were drawn up: one that indicates consent to the exam, and another that indicates consent to police seizing the SAEK for analysis (Griffiths, 1999). Victims are told that they can skip parts of the exam (except for the DNA buccal swap, which is required for all SAEKs) and stop it at any point once it has begun. However, according to many rape crisis advocates, these choices that victims make can be later used as evidence in police investigations and criminal trials that they are not being truthful.

22 CFS, "Hospital Instructions" (2012), obtained through FOI request to the Centre of Forensic Sciences, no. CSCS-A-2012–03930.

23 The booklet that accompanied the first SAEK in 1979 warned medicolegal practitioners that, "individuals in crisis are severely disorganized and have difficulty making decisions" (Provincial Secretariat for Justice, 1979a, p. 52). This understanding of women who experience sexual assault is clearly still firmly entrenched in medical and legal practice.

24 For more discussion on how police and defence lawyers use kits to corroborate and challenge victims' reports of sexual assault in the United States and the United Kingdom see Corrigan (2013a), Bumiller (2008), and Rees (2012).

25 When other exhibits were included, such as the victim's clothing, the percentage rose to 50%.

26 There are exceptions. Children and youth under the age of 16 (provided the accused is more than five years older or in a position of trust) cannot legally consent to sexual activity. DNA evidence in these cases is used to confirm the identity of the perpetrator and demonstrate that sexual activity took place.

27 Existing research on victims' varied experiences with the SAEK suggests that this dichotomy between eager victims wanting the SAEK exam and reluctant false complainants who do not is a vast oversimplification (Du Mont, White, & McGregor, 2009; Mulla, 2014). It appears particularly simplistic in the face of the many constraints and barriers that are placed on a victim's decision to have a SAEK exam discussed earlier in this chapter.

28 This strategy, of course, hinges on the assumption that *real* narratives of trauma are consistently told, regardless of who is listening, where, when, for what purpose, and under what conditions. Available evidence

on victims' experiences of reporting their assaults suggests that this is a completely inaccurate assumption (Skinner & Taylor, 2009).

6 Reassembling Technoscience

1 Jane Doe (2003) recounts the details of her story in her book *The Story of Jane Doe: A Book about Rape*. Also, see Hodgson (2010) for a case study of the Metropolitan Toronto police response to Jane Doe's rape and the systemic discrimination it revealed in policing practice.
2 For more discussion on the legal precedents that Jane Doe's case set, see Sheehy (2012) and Dewart (2012).
3 The short time frames in which these samples were taken likely had a significant effect on the rates of DNA recovery.
4 In the United States (as of 3 June 2013 [Maryland v. King, 2013]) and in the United Kingdom, police can legally obtain DNA samples from criminal suspects upon arrest. In Canada, obtaining DNA samples without a warrant or court order is currently prohibited under section 8 of the Canadian Charter of Rights and Freedoms. With the recent change in American law, future Canadian Charter challenges around DNA testing may be possible.

References

Agencies split over aid for rape victims. (1984, February 2). [Newspaper clipping from *Hamilton Spectator*]. Subject files of Provincial Secretariat for Justice, 1974–85. RG 64-4, file "Rape crisis centre." Archives of Ontario, Toronto.

Aiken, M. M., & Speck, P. M. (1995, October). Sexual assault and multiple trauma: A sexual assault nurse examiner (SANE) challenge. *Journal of Emergency Nursing, 21*(5), 466–468. Medline:7500583

Akrich, M. (1992). The de-scription of technical objects. In W. Bijker & J. Law (Eds.), *Shaping technology / building society* (pp. 205–224). Cambridge, MA: MIT Press.

Alderden, M. A., & Ullman, S. E. (2012, January). Gender difference or indifference? Detective decision making in sexual assault cases. *Journal of Interpersonal Violence, 27*(1), 3–22. http://dx.doi.org/10.1177/0886260511416465. Medline:21810790

Allan, B., & Smylie, J. (2015). *First Peoples, second class treatment: The role of racism in the health and well-being of Indigenous peoples in Canada.* Toronto : The Wellesley Institute.

Allard, S. A. (1997). Rethinking battered woman syndrome: A black feminist perspective. In K. J. Maschke (Ed.), *The legal response to violence against women* (pp. 73–90). New York: Garland Publishing, Inc.

Annual report. (1988–9). [Annual report for Women's College Hospital]. The Miss Margaret Robins Archives of Women's College Hospital, Toronto.

Aronson, J. (2007). *Genetic witness: Science, law, and controversy in the making of DNA profiling.* New Brunswick, NJ: Rutgers University Press.

Auditor General of Canada. (2007). Management of Forensic Laboratory Services: Royal Canadian Mounted Police. Ottawa: Office of the Auditor General of Canada.

Auditor General of Ontario. (2007). *Centre of Forensic Sciences*. Toronto: Office of the Auditor General of Ontario.

Auditor General of Ontario (2009). *Centre of Forensic Sciences: Follow-up on VFM section 3.02, 2007 annual report*. Retrieved from http://www.auditor. on.ca/en/content/annualreports/arreports/en09/402en09.pdf

Backhouse, C. (2008). *Carnal crimes: Sexual assault law in Canada 1900–1975*. Toronto: Irwin Law.

Baeza, J., & Turvey, B. (2002). False reports. In B. Turvey (Ed.), *Criminal profiling: An introduction to behavioural evidence analysis* (pp. 169–187). London: Academic Press.

Baker, A. (1983, July 21). Special centres proposed to treat assaulted women. *Globe and Mail*, p. 5.

Bannerji, H., Brand, D., Khosla, P., & Silvera, M. (1983). Conversation. *Fireweed: A Feminist Quarterly, 16*, 1–15.

Barnhouse, D. (1998, March 2). [Letter to Patricia Campbell]. Nancy Malcolm Fonds (MAL, file 1–1-36). The Miss Margaret Robins Archives of Women's College Hospital, Toronto.

Bass, A. (1987, December 2). DNA "fingerprint" powerful police tool. *The Gazette*, p. A1.

Baziuk, L. (2011, June). Audit report: RCMP backlog getting worse. *National Post*. Retrieved from http://news.nationalpost.com/news/canada/audit-report-rcmp-backlog-getting-worse

Beres, M., Crow, B., & Gotell, L. (2009). The perils of institutionalization of neoliberal times: Results of a national survey of Canadian sexual assault and rape crisis centres. *Canadian Journal of Sociology, 34*(1), 135–163.

Bernardo DNA sample sat on shelf 25 months before test completed. (1996, July 12). *Ottawa Citizen*, p. A4.

Bieber, F. (2004). Science and technology of forensic DNA profiling: Current use and future directions. In D. Lazer (Ed.), *DNA and the criminal justice system: The technology of justice* (pp. 23–62). London: MIT Press.

Biggs, M., Stermac, L. E., & Divinsky, M. (1998, July 14). Genital injuries following sexual assault of women with and without prior sexual intercourse experience. *Canadian Medical Association Journal, 159*(1), 33–37. Medline:9679484

Bijker, W. (1997). *Of bicycles, bakelites, and bulbs: Toward a theory of sociotechnical change*. Cambridge, MA: MIT Press.

Bill C-127. (1982). *An act to amend the Criminal Code in relation to sexual offences and other offences against the person and to amend certain other acts in relation thereto or in consequence thereof*. 1st Sess. 32nd Parl. (assented to August 4, 1982).

Bonokoski, M. (2012, August 25). McGuinty's cruelest cut of all: Elimination of Centre of Forensic Sciences' electronic unit borders on criminal. *Toronto Sun*. Retrieved from http://www.torontosun.com

Brean, J. (2016, February 2). How Jian Gohmeshi's lawyer was able to navigate Canada's rape shield law. *National Post* online. Retrieved from http://news.nationalpost.com/news/canada/how-jian-ghomeshis-lawyer-was-able-to-breach-canadas-rape-shield-law

Breitkreuz, G. (2009). [Standing Committee on Public Safety and National Security] *Statutory review of the DNA Identification Act*. Ottawa: House of Commons Canada.

Brennan, S., & Taylor-Butts, A. (2008). *Sexual assault in Canada*. Catalogue 85F0033M, no. 19. Ottawa: Canadian Centre for Justice Statistics Profile Series, Statistics Canada.

Brief submitted by: Rape crisis centre. (1974). [Report]. Canadian Women's Movement Archives Fonds, X10-1, box 106, folder "Toronto rape crisis centre (Toronto, ON): Facts, figures, briefs and other resource material [after 1970]–1990 (1 of 3)." University of Ottawa Archives and Special Collections, Ottawa.

Brockman, J. (2001). *Gender in the legal profession: Fitting or breaking the mould*. Vancouver: University of British Columbia Press.

Brodie, A. (1987, March 17). Montfort hospital proposed for sexual assault centre. *Ottawa Citizen*, p. C1.

Brown, B., Burman, M., & Jamieson, L. (1993). *Sex crimes on trial: The use of sexual evidence in Scottish courts*. Edinburgh: Edinburgh University Press.

Brownmiller, S. (1975). *Against our will: Men, women, and rape*. Toronto: Bantam Books.

Bumiller, K. (2008). *In abusive state: How neoliberalism appropriated the feminist movement against sexual violence*. London: Duke University Press.

Burgess, A. W., & Holmstrom, L. L. (1973, October). The rape victim in the emergency ward. *American Journal of Nursing, 73*(10), 1740–1745. Medline:4491033

Burman, M., Jamieson, L., Nicholson, J., & Brooks, O. (2007). Impact of aspects of the law of evidence in sexual offence trials: An evaluation study – research findings. Retrieved from http://www.gov.scot/Resource/Doc/197710/0052889.pdf

Burnett, S. (2007, July). Community caring: One-stop care for DV. *Network News: Ontario Network of Sexual Assault / Domestic Violence Treatment Centres*.

Cahill, A. (2001). *Rethinking rape*. London: Cornell University Press.

Callon, M. (1999). Actor-Network Theory: The market test. In J. Law & J. Hassard (Eds.), *Actor Network Theory and after* (pp. 181–195). Oxford; Malden, MA: Blackwell.

Callon, M., & Law, J. (1997). After the individual in society: Lessons on collectivity from science, technology, and society. *Canadian Journal of Sociology, 22*(2), 165–182. http://dx.doi.org/10.2307/3341747

Campbell, A. (1996). *Bernardo investigation review*. Ottawa: Ministry of the Attorney General.

Campbell, R., Patterson, D., Bybee, D., & Dworkin, E. (2009). Predicting sexual assault prosecution outcomes: The role of medical forensic evidence collected by sexual assault nurse examiners. *Criminal Justice and Behavior, 36*(7), 712–727. http://dx.doi.org/10.1177/0093854809335054

Campbell, R., Patterson, D., & Bybee, D. (2012, February). Prosecution of adult sexual assault cases: A longitudinal analysis of the impact of a sexual assault nurse examiner program. *Violence Against Women, 18*(2), 223–244. http://dx.doi.org/10.1177/1077801212440158. Medline:22433229

Campbell, R., Patterson, D., & Fehler-Cabral, G. (2010, December). Using ecological theory to evaluate the effectiveness of an indigenous community intervention: A study of Sexual Assault Nurse Examiner (SANE) programs. *American Journal of Community Psychology, 46*(3–4), 263–276. http://dx.doi.org/10.1007/s10464-010-9339-4. Medline:20853158

Campbell, R., Patterson, D., & Lichty, L. F. (2005, October). The effectiveness of sexual assault nurse examiner (SANE) programs: A review of psychological, medical, legal, and community outcomes. *Trauma, Violence & Abuse, 6*(4), 313–329. http://dx.doi.org/10.1177/1524838005280328. Medline:16217119

Campbell, R., & Raja, S. (1999, Fall). Secondary victimization of rape victims: Insights from mental health professionals who treat survivors of violence. *Violence and Victims, 14*(3), 261–275. Medline:10606433

Campling, C. (1980, March 25). [Letter to Gord Walker]. Records of the Provincial Secretariat for Justice special project on helping victims of sexual assault. RG 64-10, file "Sexual assault – victims." Archives of Ontario, Toronto.

Canadian Association of Chiefs of Police (2011). *Resolutions adopted at the 106th annual conference*. Kanata, ON: Canadian Association of Chiefs of Police.

Canadian Rape Crisis Centres (1979). [Funding manual]. Canadian Women's Movement Archives Fonds, X10-1, box 12, folder "Canadian rape crisis centres [Ottawa]: A funding manual for rape crisis centres, 1979." University of Ottawa Archives and Special Collections, Ottawa.

Casper, M. J., & Clarke, A. E. (1998, April). Making the Pap smear into the "right tool" for the job. *Social Studies of Science, 28*(2), 255–290. http://dx.doi.org/10.1177/030631298028002003. Medline:11620085

Caufield, N. (2003, April). Association cues: Focus on technique. *Network News: Ontario Network of Sexual Assault / Domestic Violence Treatment Centres.*

CBC (Producer). (1971). Rape in court. [Radio broadcast]. Toronto.

CBC (Producer). (2013). Rape kit backlog. [Radio broadcast]. Toronto.

CBC (2012, May 25). RCMP to close labs in Halifax, Winnipeg, Regina. *CBC News* online. Retrieved from http://www.cbc.ca/news/politics/rcmp-to-close-labs-in-halifax-winnipeg-regina-1.1186404

CBC (2015, March 25). Rural Nova Scotia lagging in sexual assault care. *CBC News* online. Retrieved from http://www.cbc.ca/news/canada/nova-scotia/rural-nova-scotia-lagging-in-sexual-assault-care-1.3009832

Centre of Forensic Sciences (2011). *Quarterly performance report: For the period October 1 to December 31, 2011.* Toronto: Centre of Forensic Sciences.

Chandler, J. (2010). "Science discovers, genius invents, industry applies, and man adapts himself ...": Some thoughts on human autonomy, law and technology. *Bulletin of Science, Technology & Society, 30*(1), 14–17. http://dx.doi.org/10.1177/0270467609355049

Chaddock, A. (January 15, 1979) [Letter to Commissioner, Ministry of the Solicitor General]. OPP Crime Prevention Program. Fonds RG 23, Series G-3, "OPP Crime Prevention Program, box 14, file "Sexual Assault – victims." Archives of Ontario, Toronto.

Clark, L., & Lewis, D. (1977). *Rape: The price of coercive sexuality.* Toronto: The Women's Press.

Clarke, A., & Montini, T. (1993, Winter). The many faces of RU486: Tales of situated knowledges and technological contestations. *Science, Technology & Human Values, 18*(1), 42–78. http://dx.doi.org/10.1177/016224399301800104. Medline:11652075

Cohen, J. E. (1990, February). DNA fingerprinting for forensic identification: Potential effects on data interpretation of subpopulation heterogeneity and band number variability. *American Journal of Human Genetics, 46*(2), 358–368. Medline:2301401

Cohen, M. (1993). The Canadian women's movement. In R. R. Pierson, M. G. Cohen, P. Bourne, & P. Masters (Eds.), *Canadian women's issues. Strong voices: Twenty-five years in women's activism in English Canada* (pp. 1–27). Toronto: James Lorimer & Company Ltd.

Colby, C. (2008, January). Acknowledge/celebrate. *Network News: Ontario Network of Sexual Assault / Domestic Violence Treatment Centres.*

Cole, S. (1989, August/September). The sex factor in violence. *Broadside, 10*(5), 12–13.

Cole, S. (2001). *Suspect identities: A history of fingerprinting and criminal identification.* Cambridge, MA: Harvard University Press.

Cole, S., & Lynch, M. (2006). The social and legal construction of suspects. *Annual Review of Law and Social Science, 2*(1), 39–60. http://dx.doi.org/10.1146/annurev.lawsocsci.2.081805.110001

Committee Against Rape and Sexual Assault. (1980). *President's annual report.* Niagara, ON: Committee Against Rape and Sexual Assault.

[Consent form]. [ca. 1981]. [Consent form for SAEK]. Records of the Provincial Secretariat for Justice special project on helping victims of sexual assault. RG 64-10, file "Sexual assault – victims." Archives of Ontario, Toronto.

Cornell, D. H. (1998, February). Helping victims of rape: A program called SANE. *New Jersey Medicine, 2,* 45–46. Medline:9505508

Cornish, R. (1982, December 19). [Letter to emergency room physicians, nurses, police forces, Crown attorneys]. John R. Haslehurst Fonds, HAS, file 73, item 33. The Miss Margaret Robins Archives of Women's College Hospital, Toronto.

Corrigan, R. (2013a). *Up against a wall: Rape reform and the failure of success.* New York: New York University Press. http://dx.doi.org/10.18574/nyu/9780814707937.001.0001.

Corrigan, R. (2013b). The new trial by ordeal: Rape kits, police practices, and the unintended effects of policy innovation. *Law & Social Inquiry, 38*(4), 920–949. http://dx.doi.org/10.1111/lsi.12002

Corroborating charges of rape. (1967). *Columbia Law Review, 67,* 1137–1138.

Counsell, M. T. (2007, May 9). *Senate speech regarding Bill C-18, Bill to amend certain acts in relation to DNA Identification, Ottawa, ON.* Senate speech, Ottawa.

Crawford, T. (1984, September 22). Sample kit credited with catching rapists. *Toronto Star,* p. A13.

Crew, A.B. (2012). Striking back: The viability of a civil action against the police for the "wrongful unfounding" of reported rapes. In E. Sheehy (Ed.), *Sexual assault in Canada: Law, legal practice and women's activism* (pp. 211–239). Ottawa: University of Ottawa Press.

Curran, T. (1997). *Forensic DNA analysis: Technology and applications.* (Report no. BP-443E). Ottawa: Parliamentary Research Branch.

Daemmrich, A. (1998). The evidence does not speak for itself: Expert witnesses and the organization of DNA-typing companies. *Social Studies of Science, 28*(5–6), 741–772. http://dx.doi.org/10.1177/030631298028005004

Dandino-Abbott, D. (1999, August). Birth of a sexual assault response team: The first year of the Lucas County / Toledo, Ohio, SART program. *Journal of Emergency Nursing, 25*(4), 333–336. http://dx.doi.org/10.1016/S0099-1767(99)70065-6. Medline:10424967

Davis, A. (1983). *Women, race, & class.* New York: Vintage.

Davis, F. (1999). *Moving the mountain: The women's movement in America.* Chicago: University of Illinois Press.

Dellinger Page, A. (2010). True colours: Police officers and rape myth acceptance. *Feminist Criminology, 5*(4), 315–334. http://dx.doi.org/10.1177/1557085110384108

Dempsey, M. (2009, January). Vicarious trauma. *Network News: Ontario Network of Sexual Assault / Domestic Violence Treatment Centres.*

Dewart, S. (2012). Jane Doe v Toronto Commissions of Police: A view from the bar. In E. Sheehy (Ed.), *Sexual assault in Canada: Law, legal practice and women's activism* (pp. 47–52). Ottawa: University of Ottawa Press.

Dionne, M. (1997). Voices of women not heard: The Bernardo investigation review. Report of Mr. Justice Archie Campbell. *Canadian Journal of Women and the Law, 9,* 394–417.

DNA Identification Act. S.C. c. 37 (1998).

DNA prints big advance. (1987, December 10). *Ottawa Citizen,* p. H8.

Doe, J. (2003). *The story of Jane Doe.* Toronto: Vintage Canada.

Doe, J. (2012). Who benefits from the sexual assault evidence kit? In E. Sheehy (Ed.), *Sexual assault in Canada: Law, legal practice and women's activism* (pp. 355–388). Ottawa: University of Ottawa Press.

Doe, J. (2016, February 18). Is Jian Ghomeshi guilty or innocent? Sadly it may not matter. *Now Toronto* online. Retrieved from https://nowtoronto.com/news/is-jian-ghomeshi-guilty-or-innocent-sadly-it-may-not-matter/

Donadio, B., & White, M.A. (1974, April). Seven who were raped. *Nursing Outlook, 22*(4), 245–247. Medline:4493356

Doolittle, R. (2017, February 3). Why police dismiss 1 in 5 sexual assault claims as baseless. *Globe and Mail.* Retrieved from http://www.theglobeandmail.com/news/investigations/unfounded-sexual-assault-canada-main/article33891309/

Dubinsky, K. (1993). *Improper advances: Rape and heterosexual conflict in Ontario 1880–1929.* Chicago: University of Chicago Press.

Dugdale, A. (2000). Intrauterine conception devices, situated knowledges, and making of women's bodies. *Australian Feminist Studies, 15*(32), 165–176. http://dx.doi.org/10.1080/08164640050138680

Du Mont, J., McGregor, M. J., Myhr, T. L., & Miller, K.-L. (2000). Predicting legal outcomes from medicolegal findings: An examination of sexual assault in two jurisdictions. *Journal of Women's Health and Law, 1*(3), 219–233.

Du Mont, J., Miller, K.-L., & Myhr, T. (2003). The role of "real rape" and "real victim" stereotypes in the police reporting practices of sexually assaulted women. *Violence Against Women, 9*(4), 466–486. http://dx.doi.org/10.1177/1077801202250960

Du Mont, J., & Parnis, D. (2001). Constructing bodily evidence through sexual assault evidence kits. *Griffith Law Review, 10,* 63–76.

Du Mont, J., & Parnis, D. (2003, August). Forensic nursing in the context of sexual assault: Comparing the opinions and practices of nurse examiners and nurses. *Applied Nursing Research, 16*(3), 173–183.http://dx.doi.org/ 10.1016/S0897-1897(03)00044-2. Medline:12931331

Du Mont, J., & White, D. (2007). *The uses and impacts of medico-legal evidence in sexual assault cases: A global review.* Sexual Violence Research Initiative. Retrieved from http://apps.who.int/iris/bitstream/10665/43795/1/9789241596046_eng.pdf

Du Mont, J., White, D., & McGregor, M. J. (2009, February). Investigating the medical forensic examination from the perspectives of sexually assaulted women. *Social Science & Medicine, 68*(4), 774–780. http://dx.doi.org/10.1016/j.socscimed.2008.11.010. Medline:19095341

Dunlop, M. (1987, April 20). Genetic "fingerprints" can identify us. *Windsor Star*, p. C1.

Eckert, W. (1978). Medical/legal and forensic services from a forensic scientist's point of view. *Journal of the Kansas Bar Association, 47*, 17–28.

Ericksen, J., Dudley, C., McIntosh, G., Ritch, L., Shumay, S., & Simpson, M. (2002, February). Clients' experiences with a specialized sexual assault service. *Journal of Emergency Nursing, 28*(1), 86–90. http://dx.doi.org/10.1067/men.2002.121740. Medline:11830744

Estrich, S. (1986). Rape. *Yale Law Journal, 95*(6), 1087–1184. http://dx.doi.org/10.2307/796522

Epstein, S. (1996). *Impure science: AIDS, activism, and the politics of knowledge.* Berkeley: University of California Press.

Evrard, J. R. (1971, September 15). Rape: The medical, social, and legal implications. *American Journal of Obstetrics and Gynecology, 111*(2), 197–199. http://dx.doi.org/10.1016/0002-9378(71)90889-1. Medline:5098588

Executive Committee Minutes. (1983, May 24). [Women's College Hospital executive meeting minutes]. John R. Haslehurst Fonds, HAS, file 73, item 15. The Miss Margaret Robins Archives of Women's College Hospital, Toronto.

Fahrney, P. (1974). Sexual assault package: A refinement of a previous idea. *Emergency Methods and Techniques, 4*(4), 340–341.

Farmen, R. K., Haukeli, I., Ruoff, P., & Frøyland, E. S. (2012, October). Assessing the presence of female DNA on post-coital penile swabs: Relevance to the investigation of sexual assault. *Journal of Forensic and Legal Medicine, 19*(7), 386–389. http://dx.doi.org/10.1016/j.jflm.2012.02.029. Medline:22920760

Feldberg, G. (1997). Defining the facts of rape: The uses of medical evidence in sexual assault trials. *Canadian Journal of Women and the Law, 9*(1), 89–114.

Feldberg, G., Ladd-Taylor, M., Li, A., & McPherson, K. (2003). *Women, health, and nation: Canada and the United States since 1945.* London: McGill-Queen's University Press.

Fitzgerald, K. (2006, April). DFSA study: "Not in our community!" *Network News: Ontario Network of Sexual Assault / Domestic Violence Treatment Centres.*

Fitzgerald, K., & Rioch, L. A. (2009, October). Kenora's RNs and RPNs: A collaborative nursing team. *Network News: Ontario Network of Sexual Assault / Domestic Violence Treatment Centres.*

Fitzgerald, M. (1982, November). Vancouver rape relief: Frustrations with the fortress mentality. *Broadside, 4*(2), 4.

Fitzpatrick, M., Ta, A., Lenchus, J., Arheart, K. L., Rosen, L. F., & Birnbach, D. J. (2012, January). Sexual assault forensic examiners' training and assessment using simulation technology. *Journal of Emergency Nursing, 38*(1), 85–90.e6. http://dx.doi.org/10.1016/j.jen.2010.10.002. Medline:22226138

Flanagan, N., & McAlister, C. (2011, November). The transfer and persistence of DNA under the fingernails following digital penetration of the vagina. *Forensic Science International. Genetics, 5*(5), 479–483. http://dx.doi.org/10.1016/j.fsigen.2010.10.008. Medline:21056024

Flood, M., & Pease, B. (2009, April). Factors influencing attitudes to violence against women. *Trauma, Violence & Abuse, 10*(2), 125–142. http://dx.doi.org/10.1177/1524838009334131. Medline:19383630

Forensic evidence form. [ca. 1981]. [SAEK forensic evidence form]. Records of the Provincial Secretariat for Justice special project on helping victims of sexual assault. RG 64-10, file "Sexual Assault – victims." Archives of Ontario, Toronto.

Forensic Nurses' Society of Canada. (2012, November 1). Forensic Nurses' Society of Canada [Website]. Retrieved from http://www.forensicnurse.ca/

Fraser, L. (2016, February 14). Jian Ghomeshi trial: Complainant in different case recounts "trauma of testifying." *CBC News* online. Retrieved from http://www.cbc.ca/news/canada/toronto/ghomeshi-mandi-gray-1.3446672

Frederiksen, S. (2011). The National Academy of Sciences, Canadian DNA jurisprudence and changing forensic practice. *Manitoba Law Journal, 35*(1), 111–142.

Friedan, B. (1963). *The feminine mystique.* New York: Bantam Dell.

Fujimura, J. (1991). On methods, ontologies, and representation in the sociology of science: Where do we stand? In D. Maines (Ed.), *Social organization and social processes: Essays in honour of Anselm L. Strauss* (pp. 207–248). Hawthorne, NY: Aldine de Gruyter.

Furst, S. (2005, July). Changes abound! Centre focus. *Network News: Ontario Network of Sexual Assault / Domestic Violence Treatment Centres.*

Gavey, N. (2005). *Just sex? The cultural scaffolding of rape.* London, Eng.: Routledge.

General examination (ca. 1981). [Body map]. Records of the Provincial Secretariat for Justice special project on helping victims of sexual assault. RG 64-10, file "Sexual Assault – victims." Archives of Ontario, Toronto.

Gerlach, N. (2004). *The genetic imaginary: DNA in the Canadian criminal justice system*. Toronto: University of Toronto Press.

Gieryn, T. (1983). Boundary-work and the demarcation of science from non-science: Strains and interests in professional ideologies of scientists. *American Sociological Review, 48*(6), 781–795.

Gill, P., Jeffreys, A. J., & Werrett, D. J. (1985, December 12–18). Forensic application of DNA "fingerprints." *Nature, 318*(6046), 577–579. http://dx.doi.org/10.1038/318577a0. Medline:3840867

Gingras, F., Paquet, C., Bazinet, M., Granger, D., Marcoux-Legault, K., Fiorillo, C., et al. (2009). Biological and DNA evidence in 1000 sexual assault cases. *Forensic Science International. Genetics Supplement Series, 2*(1), 138–140. http://dx.doi.org/10.1016/j.fsigss.2009.09.006

Griffiths, J. (1999). *Review of the investigation of sexual assault. Toronto Police Service*. Toronto: Toronto Audit Services.

Halfon, S. (1998). Collecting, testing, and convincing: Forensic DNA experts in the courts. *Social Studies of Science, 28*(5–6), 801–828. http://dx.doi.org/10.1177/030631298028005006

Halfon, S. (2010). Encountering birth: Negotiating expertise, networks, and my STS self. *Science and Culture, 19*(1), 61–77. http://dx.doi.org/10.1080/09505430903558062

Haraway, D. (1994). A game of cat's cradle: Science studies, feminist theory, cultural studies. *Configurations, 2*(1), 59–71. http://dx.doi.org/10.1353/con.1994.0009

Haraway, D. (1997). *Modest_witness@second_millennium.femaleman©_meets_oncomouse*. New York: Routledge.

Haraway, D. (2000). *How like a leaf: An interview with Thyrza Nichols Goodeve*. New York: Routledge.

Harding, S. (1991). *Whose science? Whose knowledge: Thinking from women's lives*. New York: Cornell University.

Harding, S. (2008). *Sciences from below: Feminism, postcolonialities, and modernities*. Durham: Duke University Press.

Hargot, L. (1982). Sexual assault regional centre. *Canadian Association of Emergency Physicians Review, 2*(2), 126–130.

Hargot, L. (1985). Dr. Hargot replies. *Canadian Family Physician / Médecin de Famille Canadien, 31*, 1453.

Harris, A. (1990). Race and essentialism in feminist legal theory. *Stanford Law Review, 42*(3), 581–616. http://dx.doi.org/10.2307/1228886

Hasson, K. A. (2012, Oct). Making appropriation "stick": Stabilizing politics in an "inherently feminist" tool. *Social Studies of Science, 42*(5), 638–661. http://dx.doi.org/10.1177/0306312712440750. Medline:23189608

Hatmaker, D., Pinholster, L., Saye, J. (2002). A community-based approach to sexual assault. *Public Health Nursing, 19*(2), 124–127.

Health Care Consent Act, S.O. 1996, c. 2, Schedule A (1996).

Hepworth, D., & McLeod, G. (2005). RCMP Forensic Laboratory Services: "Truth, through science and integrity." Retrieved from http://www.cssa-cila.org/garryb/publications/JusticeCommitteeBrief_HepworthMcleod_may2005.doc

Herbert, C. P., & Whynot, E. M. (1985, August). Sex assault exam: Too complicated? *Canadian Family Physician / Médecin de Famille Canadien, 31,* 1453. Medline:21274028

Hilts, P. (1987, September 22). DNA makes criminal identification certain. *Washington Post,* C16.

Hodgson, J. (2010). Policing sexual violence: A case study of Jane Doe v. the Metropolitan Toronto Police. In J. Hodgson & D. Kelley (Eds.), *Sexual violence: Policies, practices, and challenges in the United States and Canada* (pp. 173–189). London: Lynne Rienner Publishers.

Hoeffel, J. (1990). The dark side of DNA profiling: Unreliable scientific evidence meets the criminal defendant. *Stanford Law Review, 42*(2), 465–538. http://dx.doi.org/10.2307/1228965

Hoffman, L. (1989). *The politics of knowledge: Activist movements in medicine and planning.* Albany: State University of New York Press.

Holmes, H. (1994). DNA fingerprints and rape: A feminist assessment. *Policy Sciences, 27*(2–3), 221–245. http://dx.doi.org/10.1007/BF00999889

Hudlow, W. R., & Buoncristiani, M. R. (2012, January). Development of a rapid, 96-well alkaline based differential DNA extraction method for sexual assault evidence. *Forensic Science International. Genetics, 6*(1), 1–16. http://dx.doi.org/10.1016/j.fsigen.2010.12.015. Medline:21288791

Hughes, E. (1971). *The sociological eye: Selected papers.* New Brunswick, NJ: Transaction Publishers.

Human Rights Watch. (2009). *Testing justice: The rape kit backlog in Los Angeles City and County.* New York: Human Rights Watch.

Hutson, L. A. (2002, March). Development of sexual assault nurse examiner programs.*The Nursing Clinics of North America, 37*(1), 79–88. Medline:11818263

Idhe, D. (2002). *Bodies in technology.* London: University of Minnesota Press.

In brief: Forensic kits issued for sex evidence. (1981, January 16). *Globe and Mail,* p. P5.

Interviews with women's health experts. (1999). [Article downloaded from Internet]. Strategic Communications Fonds (D19, file 32). The Miss Margaret Robins Archives of Women's College Hospital, Toronto.

Irving, T. (2008). Decoding black women: Policing practices and rape prosecution on the streets of Philadelphia. *NWSA Journal, 20*(2), 100–120.

Jail term cut for girls' killer. (2009, May 14). *BBC News.* Retrieved from http://news.bbc.co.uk

Jasanoff, S. (1995). *Science at the bar: Law, science, and technology in America.* London: Harvard University Press.

Jasanoff, S. (1998). The eye of everyman: Witnessing DNA in the Simpson trial. *Social Studies of Science, 28*(5–6), 713–740. http://dx.doi.org/10.1177/030631298028005003

Jasanoff, S. (2006, Summer). Just evidence: The limits of science in the legal process. *Journal of Law, Medicine & Ethics, 34*(2), 328–341. http://dx.doi.org/10.1111/j.1748-720X.2006.00038.x. Medline:16789954

Jeffreys, A. J., Wilson, V., & Thein, S. L. (1985). Hypervariable "minisatellite" regions in human DNA. *Nature, 314*, 67–73.

Johnson, D. (2010). Sorting out the question of feminist technology. In L. Layne, S. Vostral, & K. Boyer (Eds.), *Feminist technology* (pp. 36–54). Chicago: University of Illinois Press.

Johnson, D., Peterson, J., Sommers, I., & Baskin, D. (2012, February). Use of forensic science in investigating crimes of sexual violence: Contrasting its theoretical potential with empirical realities. *Violence Against Women, 18*(2), 193–222. http://dx.doi.org/10.1177/1077801212440157. Medline:22433228

Johnson, H. (2012). Limits of a criminal justice response: Trends in police and court processing of sexual assault. In E. Sheehy (Ed.), *Sexual assault in Canada: Law, legal practice and women's activism* (pp. 613–634). Ottawa: University of Ottawa Press.

Johnson, H., & Dawson, M. (2011). *Violence against women in Canada: Research and policy perspectives.* New York: Oxford University Press.

Joyful Heart Foundation. (2015). *End the backlog.* Retrieved from http://www.endthebacklog.org

Kane, L. (2015, October). Rape kit inaccessibility a hurdle to justice for victims in Canada, say advocates. *CBC News* online. Retrieved from http://www.cbc.ca/news/canada/british-columbia/rape-kit-justice-victims-squamish-1.3267448

Kaplan, S. (2004, April). 3 Rs and the big V: Curriculum guidelines. *Network News: Ontario Network of Sexual Assault / Domestic Violence Treatment Centres.*

Keller, E. (1985). *Reflections on gender and science.* New York: Yale University Press.

Kelly, L. (1988). *Surviving sexual violence.* Minneapolis: University of Minnesota Press.

Kelly, L., Temkin, J., & Griffiths, S. (2006). *Section 41: An evaluation of new legislation limiting sexual history evidence in rape trials.* Retrieved from http://citeseerx.ist.psu.edu/viewdoc/download?doi=10.1.1.628.3925&rep=rep1&type=pdf

Kerr, G. (1978a, June 16). [Letter to Ms. Robin Jeffrey]. Records of the Provincial Secretariat for Justice special project on helping victims of sexual assault. RG 64-10, file "Rape 1978." Archives of Ontario, Toronto.

Kerr, G. (1978b). *Consultation on Rape.* Records of the Provincial Secretariat for Justice special project on helping victims of sexual assault. RG 64-10, file "Rape 1978." Archives of Ontario, Toronto.

Kerstetter, W., & Van Winkle, B. (1990). Who decides? A study of complainant's decision to prosecute in rape cases. *Criminal Justice and Behavior, 17*(3), 268–283.

Kinnon, D. (1981). *Report on sexual assault in Canada.* Ottawa: Canadian Advisory Council on the Status of Women.

Kirk, M. (1995, September 1). The Bernardo trial police had Bernardo link in 1990 physical samples not analyzed for DNA until late in 1992, after girls' deaths, sources say. *Globe and Mail,* p. A1.

Kit will catch more rapists: MD. (1981, May 13). *Toronto Star,* p. A24.

Klein, B. S. (1992, November/December). We are who you are: Feminism and disability. *Ms, 3*(3), 72.

Kling, R., & Iacono, S. (1988). The mobilization of support for computerization: Computerization movements. *Social Problems, 35*(3), 226–243. http://dx.doi.org/10.2307/800620

Kong, R., Johnson, H., Beattie, S., & Cardillo, A. (2002). Sexual offences in Canada (Catalogue no. 85-002-XIE, Vol. 23, no. 6). Ottawa: Statistics Canada.

Kozicki, D. (2007, January). A passion for the double helix: Bio-chemistry pioneer begins 34th year with RCMP. *Blue Line Magazine,* p. 42.

KPMG Consulting. (1999, December 1). *Evaluation of the domestic violence pilots: Final report.* [Report]. Nancy Malcolm Fonds. (MAL, file 2–1, item 100). The Miss Margaret Robins Archives of Women's College Hospital, Toronto.

Krishnan, S. S. (1978). *An introduction to modern criminal investigation: With basic laboratory techniques.* Springfield, IL: Charles C. Thomas Publisher.

Kubanek, J., & Miller, F. (n.d.). DNA evidence and the National DNA Databank: "Not in our name." Retrieved from http://www.casac.ca/content/dna-evidence-and-national-dna-databank-not-our-name

Lander, E. S. (1989, Jun 15). DNA fingerprinting on trial. *Nature, 339*(6225), 501–505. http://dx.doi.org/10.1038/339501a0. Medline:2567496

Lander, E. S., & Budowle, B. (1994, October 27). DNA fingerprinting dispute laid to rest. *Nature, 371*(6500), 735–738. http://dx.doi.org/10.1038/371735a0. Medline:7818670

Landsberg, M. (2011). *Writing the revolution*. Toronto: Second Story Press.

Latour, B. (1987). *Science in action: How to follow scientists and engineers through society*. Cambridge, MA: Harvard University Press.

Latour, B. (2005). *Reassembling the social: An introduction to actor-network theory*. Oxford: Oxford University Press.

Latour, B. (2010). *The making of law: An ethnography of the Conseil d'État*. Cambridge: Polity Press.

Law, J. (2002). *Aircraft stories: Decentering the object in technoscience*. Durham, NC: Duke University Press. http://dx.doi.org/10.1215/9780822383543.

Law, J. (2003). *Making a mess with method*. Retrieved from http://www.lancaster.ac.uk/fass/resources/sociology-online-papers/papers/law-making-a-mess-with-method.pdf.

Law, J. (2004). *After method: Mess in social science research*. New York: Routledge.

Law, J. (2006). Traduction/trahison: Notes on ANT. *Convergencia, 13*(42), 47–72.

Law, J. (2007, April 25). *Actor Network Theory and Material Semiotics*. Retrieved from http://www.heterogeneities.net/publications/Law2007ANTandMaterialSemiotics.pdf.

Layne, L., Vostral, S., & Boyer, K. (Eds.). (2010). *Feminist technology*. Chicago: University of Illinois Press.

LEAF (2016, March 24). Statement of solidarity with sexual assault survivors. LEAF. Retrieved from http://www.leaf.ca/statement-of-solidarity-with-sexual-assault-survivors/

LeBourdais, E. (1976, March). Rape victims: Unpopular patients. *Community Services*, 1–3.

Ledray, L. (2001). *Evidence collection and care of the sexual assault survivor: The SANE-SART response*. Violence Against Women Online Resources. Retrieved from http://citeseerx.ist.psu.edu/viewdoc/download?doi=10.1.1.644.179&rep=rep1&type=pdf.

Ledray, L. E., & Simmelink, K. (1997, February). Efficacy of SANE evidence collection: A Minnesota study. *Journal of Emergency Nursing, 23*(1), 75–77. http://dx.doi.org/10.1016/S0099-1767(97)90070-2. Medline:9128517

Lee, P. (2000). A gender critique of forensic DNA evidence: Collection, storage, and applications. Retrieved from http://www.cwhn.ca/en/node/25158

Legal-medical procedures. (1983). [Internal memo]. Canadian Women's Movement Archives Fonds, X10-1, box 106, folder "Toronto Rape Crisis Centre: Resource material, 1978–1990." University of Ottawa Archives and Special Collections, Ottawa.

Legal policy of the National Association. (1978). Canadian Women's Movement Archives Fonds, X10-1, box 12, folder "Canadian Rape Crisis Centres (Ottawa, ON): Minutes of AGM, proposals, statement and model sex offense statue, 1975–1978." University of Ottawa Archives and Special Collections, Ottawa.

L'Heureux-Dubé, C. (2001). Beyond the myths: Equality, impartiality, and justice. *Journal of Social Distress and the Homeless, 10*(1), 87–104. http:// dx.doi.org/10.1023/A:1009433703175

Lievore, D. (2005). *Prosecutorial decisions in adult sexual assault cases.* Canberra: Australian Institute of Criminology.

Light, D., & Monk-Turner, E. (2009, November). Circumstances surrounding male sexual assault and rape: Findings from the National Violence Against Women Survey. *Journal of Interpersonal Violence, 24*(11), 1849–1858. http:// dx.doi.org/10.1177/0886260508325488. Medline:18981191

Lipovenko, D. (1984, April 11). New unit offering treatment for rapes. *Globe and Mail*, M5.

Little, K. (2001, April). Sexual assault nurse examiner programs: Improving the community response to sexual assault victims. Washington, DC: United States Department of Justice. Retrieved from http://www.u.arizona. edu/~sexasslt/arpep/pdfs/186366.pdf.

Lohr, S. (1987, December 26). British police use genetic prints to convict rapist. *Windsor Star*, p. G8.

London Sexual Assault Centre. (1991, January 16). [Letter to Ministers]. Box B719579, file "Ontario Coalition of Rape Crisis Centres – Lobby January 15, 1991." Archives of Ontario, Toronto.

Louisa, T. (2010, July 7). Sexual assault victim sent to Renfrew: Critics outraged over lack of specialized service at Ottawa Hospital. *Ottawa Citizen*, p. A1.

Lynch, M., Cole, S., McNally, R., & Jordan, K. (2008). *Truth machine: The contentious history of DNA fingerprinting.* Chicago: University of Chicago Press. http://dx.doi.org/10.7208/chicago/9780226498089.001.0001.

MacCharles, T. (1988, February 3). Genetic "fingerprints" hot new crime buster. *Ottawa Citizen*, p. A1.

Macdonald, S., & Norris, P. (2010, June). *Guidelines for the collection of forensic evidence from the person who is unable to provide consent.* Toronto: Ontario Network of Sexual Assault and Domestic Violence Treatment Centres. Retrieved from http://www.satcontario.com/files/Guidelines_for_ Collection___unable_to_consent.pdf

Macdonald, S., Wyman, J., & Addison, M. (1995, June). *Guidelines and protocols for the sexual assault nurse examiner: Sexual assault care centre Women's College Hospital.* [Guidelines]. Strategic Communication Fonds (D9, file 2–41, item 2). The Miss Margaret Robins Archives of Women's College Hospital, Toronto.

MacFarlane, B. (1993). Historical development of the offence of rape. In J. Wood & R. Peck. *100 years of the Criminal Code in Canada: Essays commemorating the centenary of the Criminal Code in Canada* (pp. 1–54). Ottawa: Canadian Bar Association.

Mack, K. (1993). Continuing barriers to women's credibility: A feminist perspective on the proof process. *Criminal Law Forum, 4*(2), 327–353. http://dx.doi.org/10.1007/BF01096078

MacKinnon, C. (2005). *Women's lives, men's laws*. London: Harvard University Press.

Madigan, L., & Gamble, N. (1991). *The second rape*. New York: Lexington Books.

Maier, S. L. (2008, July). "I have heard horrible stories …": Rape victim advocates' perceptions of the revictimization of rape victims by the police and medical system. *Violence Against Women, 14*(7), 786–808. http://dx.doi.org/10.1177/1077801208320245. Medline:18559867

Maier, S. L. (2012). Sexual assault nurse examiners' perceptions of their relationship with doctors, rape victim advocates, police, and prosecutors. *Journal of Interpersonal Violence, 27*(7), 1314–1340. http://dx.doi.org/10.1177/0886260511425242. Medline:22203620

Marriner, S. (2012). Questioning "expert" knowledges. In E. Sheehy (Ed.), *Sexual assault in Canada: Law, legal practice and women's activism* (pp. 409–450). Ottawa: University of Ottawa Press.

Marshall, J. (1980, November). Attitude of doctors lets rapists go free, police officer says. *Globe and Mail*, P5.

Martin, P., DiNitto, D., Maxwell, S., & Norton, D. (1985). Controversies surrounding the rape kit exam in the 1980s: Issues and alternatives. *Crime and Delinquency, 31*(2), 223–246. http://dx.doi.org/10.1177/0011128785031002005

Martin, P., & Powell, M. (1994). Accounting for the "second assault": Legal organizations' framing of rape victims. *Law & Social Inquiry, 19*(4), 853–890.

Martin, P. Y. (2005). *Rape work: Victims, gender, and emotions in organization and community context*. New York: Routledge.

Marx, G. (1972). Introduction. In G. Marx (Ed.), *Muckracking sociology: Research as social criticism* (pp. 1–29). New Brunswick, NJ: Transaction Books.

Maryland v. King. [2013]. 569 U.S.

Masson, D. (1998). With and despite the state: Doing women's movement politics in local service groups in the 1980s in Quebec. Unpublished doctoral dissertation. Carleton University, Ottawa.

Mathieu, E., & Poisson, J. (2014a, November 20). Canadian post-secondary schools failing sex assault victims. *Toronto Star* online. Retrieved from https://www.thestar.com/news/canada/2014/11/20/canadian_postsecondary_schools_failing_sex_assault_victims.html

Mathieu, E., & Poisson, J. (2014b, November 22). Universities and colleges launch policy review. *Toronto Star* online. Retrieved from https://www.thestar.com/news/canada/2014/11/22/universities_and_colleges_launch_policy_review.html.

Maxwell, C. (2006, October). Creative integration: Community awareness. *Network News: Ontario Network of Sexual Assault / Domestic Violence Treatment Centres.*

McGregor, M. J., Du Mont, J., & Myhr, T. L. (2002, June). Sexual assault forensic medical examination: Is evidence related to successful prosecution? *Annals of Emergency Medicine, 39*(6), 639–647. http://dx.doi.org/10.1067/mem.2002.123694. Medline:12023707

McGregor, M. J., Le, G., Marion, S. A., & Wiebe, E. (1999, June 1). Examination for sexual assault: Is the documentation of physical injury associated with the laying of charges? A retrospective cohort study. *Canadian Medical Association Journal, 160*(11), 1565–1569. Medline:10373997

McMillan, L., & White, D. (2015). Minding the evidence: An exploration of sexual assault, police and forensic intervention in Scottish and Canadian contexts. Paper presented at Law and Society Association annual meeting. Seattle, WA.

Medea, A., & Thompson, K. (1975). *Against rape.* London: Peter Owen Publishers.

Meeting notes. (1983, January 31). [Notes of meeting the Attorney General called]. John R. Haslehurst Fonds, HAS, file 73, item 13. The Miss Margaret Robins Archives of Women's College Hospital, Toronto.

Meredith, C., Mohr, R., & Cairns Way, R. C. (1997). *Implementation review of Bill C-49.* Ottawa: Department of Justice.

Miller, J. (1981, December/January). Rape: Violation by the Criminal Code. *Broadside, 2*(3), 5. Retrieved from http://www.broadsidefeminist.com/images/issues/Broadside0203.pdf

Millet, K. (1969). *Sexual politics.* Chicago: University of Illinois Press.

Mills, E. (1982). One hundred years of fear: Rape and the medical profession. In N. Rafter & E. Stanko (Eds.), *Judge, lawyers, victim, thief: Women, gender roles, and criminal justice* (pp. 29–62). Boston: Northeastern University Press.

Minutes of AGM (1978). *Canadian rape crisis centres fourth annual meeting.* [Meeting minutes]. Canadian Women's Movement Archives Fonds, X10-1, box 94, folder "Canadian Rape Crisis Centres: Minutes of AGM, proposal, statement, and model sex offense statute 1975–1978 [2 of 2]." University of Ottawa Archives and Special Collections, Ottawa.

Mitchell, P. (1999). [Unnamed article]. Michele Landsberg Fonds, F250, series 1089, file 1035. Toronto: City of Toronto Archives.

Mol, A. (2002). *The body multiple: Ontology in medical practice*. Durham, NC: Duke University Press. http://dx.doi.org/10.1215/9780822384151.

Montgomery, C. (1981, April 9). Bill police for examining rape victims, MDs urged. *Globe and Mail*, P1.

Monture-Okanee, P. (1992). The violence we women do: A first nations view. In C. Backhouse & D. Flaherty (Eds.), *Challenging times: The women's movement in Canada and the United States* (pp. 193–200). Montreal: McGill-Queen's University Press.

Moore, D., & Singh, R. (2015). Seeing crime: ANT, feminism and images of violence against women. In D. Robert & M. Dufresne (Eds.), *Actor-Network Theory and Crime Studies: Explorations in science and technology* (pp. 67–80). New York: Routledge

Morgan, S. (2002). *Into our own hands: The women's health movement in the United States, 1969–1990*. Piscataway, NJ: Rutgers University Press.

Morton, M. (1980, January 14). [Letter to Alderman]. Ann Johnston Fonds, F1312, series 733, file 269. City of Toronto Archives, Toronto.

Mulla, S. (2008). There is no place like home: The body as the scene of the crime in sexual assault investigation. *Home Cultures, 5*(3), 301–326.

Mulla, S. (2014). *The violence of care: Rape victims, forensic nurses, and sexual assault intervention*. New York: New York University Press. http://dx.doi.org/10.18574/nyu/9781479800315.001.0001.

Murphy, M. (2012). *Seizing the means of reproduction: Entanglements of feminism, health, and technoscience*. London: Duke University Press. http://dx.doi.org/10.1215/9780822395805.

Murphy, M. (2016, April 5). Marie Henein: Not a feminist, not a surprise. Rabble.ca Blogs. Retrieved from http://www.feministcurrent.com/2016/04/04/marie-henein-not-a-feminist-not-a-surprise/

Murray, V. V., Jick, T. D., & Bradshaw, P. (1984). Hospital funding constraints: Strategic and tactical decision responses to sustained moderate levels of crisis in six Canadian hospitals. *Social Science & Medicine, 18*(3), 211–219. http://dx.doi.org/10.1016/0277-9536(84)90082-0. Medline:6701565

National day of protest. (1977). [Flyer]. Canadian Women's Movement Archives Fonds, X10-1, box 106, folder "Toronto Rape Crisis Centre: Facts, figures, briefs, and other resource material [after 1970–1990]." University of Ottawa Archives and Special Collections, Ottawa.

National DNA Databank. (2002). *The National DNA Databank of Canada annual report*. Ottawa: Royal Canadian Mounted Police.

National DNA Databank. (2003). *The National DNA Databank of Canada annual report*. Ottawa: Royal Canadian Mounted Police.

National DNA Databank. (2006). *The National DNA Databank of Canada annual report*. Ottawa: Royal Canadian Mounted Police.

National DNA Databank Advisory Committee. (2010). *National DNA Databank Advisory Committee Annual Report*. Ottawa: Royal Canadian Mounted Police.

National Center for Victims of Crime. (2016). *Rape kit action project*. Retrieved from https://victimsofcrime.org/our-programs/dna-resource-center/rape-kit-action-project.

Neigh, S. (2012). *Talking radical: Gender and sexuality: Canadian history through the stories of activists*. Halifax, NS: Fernwood Publishing.

Nelkin, D., & Lindee, M. S. (2004). *The DNA mystique: The gene as a cultural icon*. Ann Arbor: University of Michigan. http://dx.doi.org/10.3998/mpub.6769.

Ng, R. (1996). *The politics of community services: Immigrant women, class, and the state*. Halifax, NS: Fernwood Publishing.

[No author]. [ca. 1981]. [No title – Handwritten notes on the "Contents of the Sexual Assault Evidence Kit"]. Records of the Provincial Secretariat for Justice special project on helping victims of sexual assault. RG 64-10, file "Sexual assault – victims." Archives of Ontario, Toronto.

Odette, F. (2012). Sexual assault and disabled women ten years after Jane Doe. In E. Sheehy (Ed.), *Sexual assault in Canada: Law, legal practice and women's activism* (pp. 173–190). Ottawa: University of Ottawa Press.

One day conference. [ca. 1975]. [Conference report]. Canadian Women's Movement Archives Fonds, X10-1, box 94, folder "Rape crisis centre [Hamilton, ON]: Report of an awareness conference, national network project submission and other material, 1976–1979 [1 of 3]." University of Ottawa Archives and Special Collections, Ottawa.

Ontario Coalition of Rape Crisis Centres. (1979). [Funding application]. Records of the Provincial Secretariat for Justice special project on helping victims of sexual assault. RG 64-10, file "Sexual Assault – victims." Archives of Ontario, Toronto.

Ontario Coalition of Rape Crisis Centres. (1991). *Ontario Coalition Centres response*. [Report]. Ontario Coalition of Rape Crisis Centres 1977/98. RG 68-40, file "Ontario Coalition of Rape Crisis Centres Lobby January 15, 1991." Archives of Ontario, Toronto.

Ontario Coalition of Rape Crisis Centres. (2015). *About us: Sexual assault centres in your communities*. Ontario Coalition of Rape Crisis Centres. Retrieved from http://www.sexualassaultsupport.ca/page-411845.

Ontario Hospital Association. (1983a, January 26). *For your information*. [Newsletter from Ontario Hospital Association, vol. 17, no. 2]. Strategic Communication Fonds (D19, file 32). The Miss Margaret Robins Archives of Women's College Hospital, Toronto.

Ontario Hospital Association. (1983b, July). *The treatment of persons who have been sexually assaulted: Guidelines for hospitals.* [Guidelines]. John R. Haslehurst Fonds, HAS, file 73, item 14. The Miss Margaret Robins Archives of Women's College Hospital, Toronto.

Ontario Ministry of Health. (1982, October 5). [Letter to administrators and chiefs of medical staff at public hospitals]. John R. Haslehurst Fonds, HAS, file 73, item 27. The Miss Margaret Robins Archives of Women's College Hospital, Toronto.

Ontario Ministry of Health. [ca. 1984]. Sexual assault treatment centres to be established, Norton announces. [News release]. John R. Haslehurst Fonds, HAS, file 73, item 1. The Miss Margaret Robins Archives of Women's College Hospital, Toronto.

Ontario Women's Directorate. (1994, November). *Program review of the wife assault and sexual assault prevention initiatives.* [Report]. Michele Landsberg Fonds, F250, series 1089, file 1035. City of Toronto Archives, Toronto.

Ontario Women's Directorate. (2011). *Changing attitudes, changing lives: Ontario's sexual violence action plan.* Toronto: Government of Ontario.

OPP rape report prepared impartially. (1979, January 2). [Newspaper clipping]. OPP Crime Prevention Program. Fonds RG 23, series G-3, file "Sexual Assault – victims." Archives of Ontario, Toronto.

Osborne, J. (1984). Rape law reform: The new cosmetic for Canadian women. *Women and Politics, 4*(3), 49–64.

Parnis, D., & Du Mont, J. (1999). Rape laws and rape processing: The contradictory nature of corroboration. *Canadian Women's Studies, 19*(1), 74–78.

Parnis, D., & Du Mont, J. (2002, December). Examining the standardized application of rape kits: An exploratory study of post-sexual assault professional practices. *Health Care for Women International, 23*(8), 846–853. http://dx.doi.org/10.1080/07399330290112362. Medline:12487699

Parnis, D., & Du Mont, J. (2006). Symbolic power and the institutional response to rape: Uncovering the cultural dynamics of a forensic technology. *Canadian Review of Sociology and Anthropology / La Revue Canadienne de Sociologie et d'Anthropologie, 43*(1), 73–93. http://dx.doi.org/10.1111/j.1755-618X.2006.tb00855.x

Parnis, D., Du Mont, J., & Gombay, B. (2005, May). Cooperation or co-optation?: Assessing the methodological benefits and barriers involved in conducting qualitative research through medical institutional settings. *Qualitative Health Research, 15*(5), 686–697. http://dx.doi.org/10.1177/1049732304271832. Medline:15802543

Patterson, D., Greeson, M., & Campbell, R. (2009, May). Understanding rape survivors' decisions not to seek help from formal social systems.

Health & Social Work, 34(2), 127–136. http://dx.doi.org/10.1093/hsw/34.2.127. Medline:19425342

Pierson, R. (1993). The mainstream women's movement and the politics of difference. In R. R. Pierson, M. G. Cohen, P. Bourne, & P. Masters (Eds.), *Canadian women's issues. Strong voices: Twenty-five years in women's activism in English Canada* (pp. 186–214). Toronto: James Lorimer & Company Ltd.

Plichta, S. B., Clements, P. T., & Houseman, C. (2007, Spring). Why SANEs matter: Models of care for sexual violence victims in the emergency department. *Journal of Forensic Nursing, 3*(1), 15–23. http://dx.doi.org/10.1097/01263942-200703000-00003. Medline:17479562

Poisson, J., & Mathieu, E. (2014, November 20). Rape victim's ordeal with University of Saskatchewan. *Toronto Star* online.

Police crime statistics by offense, Ontario 1975. [1975]. [Statistical chart]. Rape Crisis Centre Funding: Wife Battering. RG 64-10, file "Rape 1978." Archives of Ontario, Toronto.

Preamble. (1975). [Minutes of annual general meeting of Canadian rape crisis centres]. Canadian Women's Movement Archives Fonds, X10-1, box 12, folder "Canadian rape crisis centres [Ottawa]: Minutes of AGM, proposal, statement, and model sex offense statute, 1975–1978 [1 of 2]." University of Ottawa Archives and Special Collections, Ottawa.

Prentice, A., Bourne, P., Brandt, G., Light, B., Mitchinson, W., & Black, N. (1996). *Canadian women: A history*. Toronto: Nelson Education.

Present laws. [ca. 1980]. [Legal summary]. Canadian Women's Movement Archives Fonds, X10-1, box 106, folder "Toronto Rape Crisis Centre facts, figures, briefs, and other resource material [after 1970] [1 of 3]." University of Ottawa Archives and Special Collections, Ottawa.

Press Progress (2016, February 11). 5 dangerous myths about sexual assault perpetuated by the Jian Ghomeshi trial. *Press Progress*. Retrieved from http://www.pressprogress.ca/5_dangerous_myths_about_sexual_assault_perpetuated_by_the_jian_ghomeshi_trial

Price, J., Gifford, J., & Summers, H. (2010). Forensic sexual abuse examination. *Paediatrics & Child Health, 20*(12), 549–555. http://dx.doi.org/10.1016/j.paed.2010.07.006

Pierson, R. R., Cohen, M. G., Bourne, P., & Masters, P. (Eds.). (1993). *Canadian women's issues: Twenty-five years of women's activism in English Canada*. Toronto: James Lorimer and Company.

Privacy Commissioner of Canada (1995). *Genetic testing and privacy*. Retrieved from http://www.priv.gc.ca/information/02_05_11_e.pdf

Procedures form. [ca. 1981]. [SAEK procedures form]. Records of the Provincial Secretariat for Justice special project on helping victims of sexual assault. RG 64-10, file "Sexual Assault – victims." Archives of Ontario, Toronto.

Provincial Secretariat for Justice (1978). *Report of a consultation on rape.* Records of the Provincial Secretariat for Justice special project on helping victims of sexual assault. RG 64-10, File "Sexual Assault – victims." Archives of Ontario, Toronto.

Provincial Secretariat for Justice. (1979a). *Helping the victims of sexual assault.* Toronto: Provincial Secretariat for Justice.

Provincial Secretariat for Justice. (1979b). *Information for the victims of sexual assault.* Toronto: Provincial Secretariat for Justice.

Provincial Secretariat for Justice (1981). *Ontario introduces standardized forensic evidence kit to help victims of sexual assault.* Records of the Provincial Secretariat for Justice special project on helping victims of sexual assault. RG 64-10, File "Sexual Assault – victims." Archives of Ontario, Toronto.

Public Affairs. (2000, November 7). *Partnerships helps victims of domestic violence.* [Memo]. Strategic Communication Fonds (D19, file 32). The Miss Margaret Robins Archives of Women's College Hospital, Toronto.

Quinlan, A. (2016). Suspect survivors: Police investigation practices in sexual assault cases in Ontario, Canada. *Women & Criminal Justice, 26*(4), 301–318. http://dx.doi.org/10.1080/08974454.2015.1124823

Quinlan, E., Quinlan, A., Fogel, C., & Taylor, G. (Eds.) (Forthcoming). Sexual violence at Canadian universities: Activism, institutional responses, and strategies for change. Waterloo, ON: Wilfrid Laurier University Press.

R. v. Baptiste. [1991]. 175 (BC SC).

R. v. Ewanchuk. [1999]. 1 SCR 330.

R. v. Ghomeshi. [2016]. ONCJ 155.

R. v. Mills. [1999]. 3 SCR 668.

R. v. Mohan. [1994]. 2 SCR 9.

R. v. O'Connor. [1995]. 4 SCR 411.

R. v. Radcliffe. [2009]. 33039 (ON SC).

R. v. S. F. [1997]. 12443 (ON SC)

R. v. Seaboyer; R. v. Gayme, [1991]. 2 SCR 577.

R. v. Thomas. [2006]. 1012 (ON SC)

Rabinow, P. (1996). *Making PCR: A story of biotechnology.* Chicago: University of Chicago Press.

Rape, Abuse, Incest National Network. (2016). *DNA and sexual violence.* Retrieved from https://www.rainn.org/dna-and-sexual-violence.

Rape crisis centre may have to close. (1979, June 21). [Clipping from *Globe and Mail*]. Canadian Women's Movement Archives Fonds, X10-1, box 94, folder "Rape Crisis Centre [Hamilton, ON]: Report of an awareness conference, national network project submission and other material, 1976–1979 [1 of 3]." University of Ottawa Archives and Special Collections, Ottawa.

Rape crisis centre opening in Toronto. [ca. 1974]. [Flyer]. Canadian Women's Movement Archives Fonds, X10-1, box 94, folder 106, "Toronto Rape Crisis Centre facts, figures, briefs, and other resource material [after 1970] [2 of 3]." University of Ottawa Archives and Special Collections, Ottawa.

RCMP offers to retest evidence following critical auditor's report. (2007, May 8). *CBC News.* Retrieved from http://www.rcmpwatch.com

Rebick, J. (2005). *Ten thousand roses: The making of a feminist revolution.* London, Eng.: Penguin.

Rees, G. (2011, September). "Morphology is a witness which doesn't lie": Diagnosis by similarity relation and analogical inference in clinical forensic medicine. *Social Science & Medicine, 73*(6), 866–872. http://dx.doi.org/10.1016/j.socscimed.2011.02.032. Medline:21440970

Rees, G. (2012). Whose credibility is it anyway: Professional authority and relevance in forensic nurse examinations of sexual assault survivors. *Review of European Studies, 4*(4), 110–120.

Rees, G. (2015). Contentious roomates? Spatial constructions of the therapeutic-evidential spectrum in medicolegal work. In I. Harper, T. Kelly, & A. Khanna (Eds.), *The clinic and the court: Law, medicine, and anthropology* (pp. 141–160). Cambridge: Cambridge University Press. http://dx.doi.org/10.1017/CBO9781139923286.007.

Relevance of rape crisis centres to the Departments of Justice and the Solicitor-General. [ca. 1977]. [Memo]. Canadian Women's Movement Archives Fonds, X10-1, box 12, folder "Canadian Rape Crisis Centres [Ottawa]: Rape crisis centre training manual for volunteers, 1977." University of Ottawa Archives and Special Collections, Ottawa.

A report of the OHA survey on hospital procedures in cases of sexual assault. (1983, January 12). [Survey results]. John R. Haslehurst Fonds, HAS, file 73, item 18. The Miss Margaret Robins Archives of Women's College Hospital, Toronto.

Report of the Royal Commission on the Status of Women. (1970). Ottawa: Information Canada. Retrieved from http://epe.lac-bac.gc.ca/100/200/301/pco-bcp/commissions-ef/bird1970-eng/bird1970-part1-eng.pdf.

Reynolds, R., Sensabaugh, G., & Blake, E. (1991). Analysis of genetic markers in forensic DNA samples using the polymerase chain reaction. *Analytical Chemistry, 1*(63), 2–15.

Riesch, H. (2010). Theorizing boundary work as representation and identity. *Journal for the Theory of Social Behaviour, 40*(4), 452–473. http://dx.doi.org/10.1111/j.1468-5914.2010.00441.x

Ringaert, L. (2003). The history of accessibility in Canada from the advocacy perspective. In D. Stienstra & A. Wight-Felske (Eds.), *Making equality: History of advocacy and persons with disabilities in Canada* (pp. 279–300). Concord, ON: Captus Press.

Robert, D., & Dufresne, M. (2015a). Thinking through networks, reaching for objects, and witnessing facticity. In D. Robert & M. Dufresne (Eds.), *Actor-network theory and crime studies: Explorations in science and technology* (pp. 1–5). New York: Routledge.

Robert, D. & Dufresne, M. (Eds.). (2015b). *Actor-network theory and crime studies: Explorations in science and technology*. New York: Routledge.

Rowland, R., & Klein, R. (1996). Radical feminism: History, politics, action. In D. Bell & R. Klein (Eds.), *Radically speaking: Feminism reclaimed* (pp. 9–36). Melbourne, Australia: Spinifex Press Pty Ltd.

Royal Canadian Mounted Police (2017, February). Statistics of National DNA Databank. Retrieved from http://www.rcmp-grc.gc.ca/nddb-bndg/stats-eng.htm

Russell, L. (2010). What women need now from police and prosecutors: 35 years of working to improve police response to male violence against women. *Canadian Women's Studies, 28*(1), 28–36.

Rutherford, A. (2011). Sexual violence against women: Putting rape research in context. *Psychology of Women Quarterly, 35*(2), 342–347. http://dx.doi.org/10.1177/0361684311404307

Sallomi, M. (2014). Coopting the antiviolence movement: Why expanding DNA surveillance won't make us safer. *Social Justice* (San Francisco), *39*(4), 97–127.

Saltmarche, A., & Cherrie, P. (2000, January). Sexual assault care centre and domestic violence: Recommendations. [Proposal]. Nancy Malcolm Fonds (MAL, file 1–1, item 36). The Miss Margaret Robins Archives of Women's College Hospital, Toronto.

Sampsel, K., Szobota, L., Joyce, D., & Graham, K., & Pickett, W. (2009). The impact of a sexual assault / domestic violence program on ED care. *Journal of Emergency Nursing, 35*(4), 282–289.

Schmitz, C. (1988, May 27). "Whack" sex assault complainant at preliminary inquiry. [Newspaper clipping]. Michele Landsberg Fonds, F250, series 1089, file 830. City of Toronto Archives, Toronto.

Scientific evidence in rape prosecution. (1980). *University of Missouri–Kansas City Law Review, 48*(2), 216–236.

Sexual Assault Centre London [ca. 1981]. *Sexual Assault Centre London*. [Pamphlet]. Canadian Women's Movement Archives Fonds, X10-1, box 102, folder "Sexual assault centre (London, ON): 1 pamphlet and London rape victims' legal handbook, 1981." University of Ottawa Archives and Special Collections, Ottawa.

Sexual Assault Crisis Centre Windsor [ca. 1984]. *Sexual Assault Crisis Centre.* [Pamphlet]. Canadian Women's Movement Archives Fonds, X10-1, box 102, folder "Sexual assault crisis centre (Windsor, ON): Pamphlets, historical overview and training manual contents, [1981]–1984." University of Ottawa Archives and Special Collections, Ottawa.

Seymour, A. (2011, August 18). Cab driver cleared of rape charge: Lack of rape kit left judge little evidence. *Ottawa Citizen,* C1.

Shapin, S., & Schaffer, S. (1985). *Leviathan and the air-pump: Hobbes, Boyle, and the experimental life.* Princeton, NJ: Princeton University Press.

Sheehy, E. (1999). Legal responses to violence against women in Canada. *Canadian Women's Studies, 19*(1), 62–73.

Sheehy, E. (2012). The many victories of Jane Doe. In E. Sheehy (Ed.), *Sexual assault in Canada: Law, legal practice and women's activism* (pp. 23–45). Ottawa: University of Ottawa Press.

Siedlikowski, H. (2004, October). Nurturing nurses: Staff retention. Network News: Ontario Network of Sexual Assault / Domestic Violence Treatment Centres.

Sievers, V., Murphy, S., & Miller, J. J. (2003, December). Sexual assault evidence collection more accurate when completed by sexual assault nurse examiners: Colorado's experience. *Journal of Emergency Medicine, 29*(6), 511–514. Medline:14631337

Sievers, V., & Stinson, S. (2002, April). Excellence in forensic practice: A clinical ladder model for recruiting and retaining sexual assault nurse examiners (SANEs). *Journal of Emergency Nursing, 28*(2), 172–175. http://dx.doi.org/10.1067/men.2002.123390. Medline:11960135

Sinclair, D. (1978). *Consultation on rape: Closing remarks and summary.* Records of the Provincial Secretariat for Justice special project on helping victims of sexual assault. RG 64-10, file "Sexual Assault – victims." Archives of Ontario, Toronto.

Sinclair, D. (1980, June 26). [Letter to Sandi Sahli]. Records of the Provincial Secretariat for Justice special project on helping victims of sexual assault. RG 64-10, file "Sexual Assault – Victims." Archives of Ontario, Toronto.

Singleton, V. (1998). Stabilizing instabilities: The role of the laboratory in the United Kingdom cervical screening programme. In M. Berg & A. Mol (Eds.), *Differences in medicine: Unraveling practices, techniques, and bodies* (pp. 86–104). London, Eng.: Duke University Press.

Skinner, T., & Taylor, H. (2009). Being shut out in the dark: Young survivors' experiences of reporting a sexual offence. *Feminist Criminology, 4*(2), 130–150. http://dx.doi.org/10.1177/1557085108326118

Slade, D. (2011, July 7). Rapist convicted 24 years later: Victim credits cold case detective, DNA. *Calgary Herald*, p. B1.

Smart, C. (1989). *Feminism and the power of law*. New York: Routledge.

Sommers, I., & Baskin, D. (2011). The influence of forensic evidence on the case ourcomes of rape incidents. *Justice System Journal, 32*(2), 314–334.

Speirs, R. (1981, October). Make convict pay: Walker. *Globe and Mail*, CL8.

Spry, T. (1995). In the absence of word and body: Hegemonic implications of "victim" and "survivor" in women's narratives of sexual violence. *Women & Language, 18*(2), 27–32.

Standing Committee on Public Accounts (2008). *Centre of Forensic Sciences: Section 3.02, 2007 annual report of the Auditor General of Ontario* (1st Session, 39th Parliament, 57 Elizabeth II). Toronto: Legislative Assembly of Ontario.

Standing Committee on Public Safety and National Security (2009). *Statutory review of the DNA Identification Act*. Ottawa: House of Commons Canada.

Star, S. L. (1989). The structure of ill-structured solutions: Boundary objects and heterogeneous distributed problem solving. In M. Huhns & L. Gasser (Eds.), *Readings in distributed artificial intelligence*, pp. 37–54. Menlo Park, CA: Kaufman.

Star, S. L. (1991). Power, technologies and the phenomenology of conventions: On being allergic to onions. In J. Law (Ed.), *A sociology of monsters? Essays on power, technology and domination* (pp. 26–56). London: Routledge.

Star, S. L. (2010). This is not a boundary object: Reflections on the origin of a concept. *Science, Technology & Human Values, 35*(5), 601–617. http://dx.doi.org/10.1177/0162243910377624

Star, S. L., & Griesemer, J. (1989). Institutional ecology, "translations" and boundary objects: Amateurs and professionals in Berkeley's museum of vertebrate zoology, 1907–1937. *Social Studies of Science, 19*(3), 387–420. http://dx.doi.org/10.1177/030631289019003001

Statement in the legislature by the Honourable Gord Walker, Q.C., Provincial Secretary for Justice. (1980, March 31). [Speech]. Provincial Secretariat for Justice special project on helping victims of sexual assault. RG 64-10, File "Sexual Assault – victims." Archives of Ontario, Toronto.

Statistics Canada. (2009). *Victimization survey, General social survey cycle 23. tabulated data through ODESI (Ontario data documentation, extraction service and infrastructure)*. Ottawa: Statistics Canada.

Stead, S. (1982, October 1). Bill made to ensure treatment in rapes. *Globe and Mail*, P1.

Stermac, L. E., & Stirpe, T. S. (2002, February). Efficacy of a 2-year-old sexual assault nurse examiner program in a Canadian hospital. *Journal of Emergency Nursing, 28*(1), 18–23. http://dx.doi.org/10.1067/men.2002.119975. Medline:11830729

Stone, L. (2014, February 3). Conservatives "streamlining" police forensics to improve on backlog. *Global News*. Retrieved from http://globalnews.ca/news/1126180/conservatives-streamlining-police-forensics-to-improve-on-backlog/

Strauss, S. (1987, September 19). "Fingerprint" genes point at the guilty. *Globe and Mail*, D4.

Stuart, D., & Delisle, R. (2004). *Learning Canadian criminal law*. Scarborough, ON: Thomson Carswell.

Suchman, L. (2003). *Located accountabilities in technology production*. Retrieved from http://www.lancaster.ac.uk/fass/resources/sociology-online-papers/papers/suchman-located-accountabilities.pdf.

Sudbury to get new sexual assault centre. (2011, December 13). *CBC News* online. Retrieved from http://www.cbc.ca/news/canada/sudbury/sudbury-to-get-new-sexual-assault-centre-1.1118504

Sudbury Regional Rape Crisis Centre [ca. 1980]. The Sudbury regional rape crisis centre. [Pamphlet]. Canadian Women's Movement Archives Fonds, X10-1, box 104, folder "Sudbury regional rape crisis centre (Sudbury, ON): 1 brochure [between 1977 and 1992]." University of Ottawa Archives and Special Collections, Ottawa.

Takeshita, C. (2012). *The global biopolitics of the IUD: How science constructs contraceptive users and women's bodies*. London: MIT Press.

Taking care: Sexual assault evidence kit training video. (1990). [Training video]. Strategic Communication Fonds (D2–094, Series WCH moving images collection). The Miss Margaret Archives of Women's College Hospital, Toronto.

Tang, S. Y., & Browne, A. J. (2008, Apr). "Race" matters: Racialization and egalitarian discourses involving Aboriginal people in the Canadian health care context. *Ethnicity & Health, 13*(2), 109–127. http://dx.doi.org/10.1080/13557850701830307. Medline:18425710

Tanovich, D., & Craig, E. (2016). Whacking the complainant: A real and current systemic problem. *Globe and Mail* online. Retrieved from http://www.theglobeandmail.com/opinion/whacking-the-complainant-is-a-real-and-current-systemic-problem/article28695366/

Tasca, M., Rodriguez, N., Spohn, C., & Koss, M. P. (2012, April). Police decision making in sexual assault cases: Predictors of suspect identification and arrest. *Journal of Interpersonal Violence, 28*(6), 1157–1177. http://dx.doi.org/10.1177/0886260512468233. Medline:23248354

Task Force on Public Violence Against Women and Children. (1983, July). [Preliminary report]. Subject files of Provincial Secretariat for Justice, 1974–1985. RG 64-4, file "Violence against women and children." Archives of Ontario, Toronto.

Taslitz, A. (1999). *Rape and the culture of the courtroom*. New York: New York University Press.

Teekah, A., Scholz, E., Friedman, M., & O'Reilly, A. (Eds.). (2015). *This is what a feminist slut looks like: Perspectives on the SlutWalk Movement*. Bradford, ON: Demeter Press.

Temkin, J. (1998). Medical evidence in rape cases: A continuing problem for criminal justice. *Modern Law Review, 61*(6), 821–848. http://dx.doi.org/10.1111/1468-2230.00180

Temkin, J. (2000). Prosecuting and defining rape: Perspectives from the bar. *Journal of Law and Society, 27*(2), 219–248. http://dx.doi.org/10.1111/1467-6478.00152

Temkin, J. (2005). *Rape and the legal process*. Oxford: Oxford University Press.

The rape corroboration requirement: Repeal not reform. (1972). *Yale Law Journal, 81*(7), 1365–1391. http://dx.doi.org/10.2307/795246

Thompson, W. (1993). Evaluating the admissibility of new genetic identification tests: Lessons from the "DNA war." *Journal of Criminal Law & Criminology, 84*(1), 22–104. http://dx.doi.org/10.2307/1143886

Timmermans, S., & Berg, M. (1997). Standardization in action: Achieving local universality through medical protocols. *Social Studies of Science, 27*(2), 273–305. http://dx.doi.org/10.1177/030631297027002003

Toppozini, D., Maxwell, C., & Mesch, H. (2003, April). Northern issues: Outreach x2. *Network News: Ontario Network of Sexual Assault / Domestic Violence Treatment Centres*.

Toronto Rape Crisis Centre (1979). *The victim of sexual assault: A handbook for the helping systems*. Canadian Women's Movement Archives Fonds, X10-1, box 106, folder "Toronto rape crisis centre bibliographies and other resource material, [1975] [1 of 2]." University of Ottawa Archives and Special Collections, Ottawa.

Toronto Rape Crisis Centre (1986). *Three alternatives to the legal system*. [Brochure]. Canadian Women's Movement Archives Fonds, X10-1, box 106, folder "Toronto rape crisis centre (Toronto, ON): Facts, figures, briefs and other resource material [after 1970]–1990 [1 of 3]." University of Ottawa Archives and Special Collections, Ottawa.

Toronto Rape Crisis Centre / Multicultural Women Against Rape (n.d.). Toronto Rape Crisis Centre / Multicultural Women Against Rape. [Website]. Retrieved from http://www.trccmwar.ca/

Training manual for volunteers. (1977). [Manual]. Canadian Women's Movement Archives Fonds, X10-1, box 12, folder "Canadian Rape Crisis Centres [Ottawa, ON]: Rape crisis centre training manual for volunteers, 1977." University of Ottawa Archives and Special Collections, Ottawa.

Ullman, S. (1996). Correlations and consequences of adult sexual assault disclosure. *Journal of Interpersonal Violence, 11*(4), 554–571. http://dx.doi.org/10.1177/088626096011004007

Ullman, S. (2010). *Talking about sexual assault: Society's response to survivors.* Washington, DC: American Psychological Association. http://dx.doi.org/10.1037/12083-000.

United States Department of Justice, Office on Violence Against Women. (2013, April). *A national protocol for sexual assault medical forensic examinations: Adults/adolescents* (NCJ 241903). Retrieved from https://www.ncjrs.gov/pdffiles1/ovw/241903.pdf.

Unknown. [ca. 1978a]. [Internal memo on implementation committee]. Records of the Provincial Secretariat for Justice special project on helping victims of sexual assault. RG 64-10, file "Rape 1978." Archives of Ontario, Toronto.

Unknown. [ca. 1978b]. *Notes for minister re CBC interview at 12:15 pm on January 20, 1978 at Queen's Park.* Provincial Secretariat for Justice special project on helping victims of sexual assault. RG 64-10, file "Rape 1978." Archives of Ontario, Toronto.

Unknown. (1986). *SACC clinical day.* [Article in House Call: Women's College Hospital, volume 6, number 2]. Strategic Communications Fonds (D19, file 32). The Miss Margaret Robins Archives of Women's College Hospital, Toronto.

Vance, J. [ca. 1977]. The experience of rape crisis centres [Unpublished article]. Canadian Women's Movement Archives Fonds, X10-1, box 12, folder "Canadian rape crisis centres (Ottawa, ON): Minutes of AGM, proposal, statement, and model sex offense statue, 1975–1978 [1 of 2]." University of Ottawa Archives and Special Collections, Ottawa.

Vance, J. (1978). *The national network of Canadian rape crisis centres: Project submission to the Department of Health and Welfare* [Funding proposal]. 1990–91/248 GAD [RG 29, box 35, file 0158-1-C2]. Library and Archives Canada, Ottawa.

van Wageningen, E. (2000, August 19). Special report. Body of evidence: Police bank on DNA to crack "cold" cases; Unsolved murders: "Very wide power" for cops. *Windsor Star*, p. A1.

Wajcman, J. (2000). Reflections on gender and Technology Studies: In what state is the art? *Social Studies of Science, 30*(3), 447–464. http://dx.doi.org/10.1177/030631200030003005

Wambaugh, J. (1989). *The blooding.* New York: Morrow.

Washington, P. (1999). Second assault of male survivors of sexual violence. *Journal of Interpersonal Violence, 14*(7), 713–730. http://dx.doi.org/10.1177/088626099014007003

Welbourn, R., & Lambert, L. (1977, December 1). *Memo to D. Sinclair re: incidence of rape.* [Internal memo]. Provincial Secretariat for Justice special project on helping victims of sexual assault. RG 64-10, file "Rape 1978." Archives of Ontario, Toronto.

White, D., & Du Mont, J. (2009, January). Visualizing sexual assault: An exploration of the use of optical technologies in the medico-legal context. *Social Science & Medicine, 68*(1), 1–8, discussion 9–11. http://dx.doi.org/10.1016/j.socscimed.2008.09.054. Medline:18952339

Why invest in rape crisis centres? (1979). [Pamphlet]. Canadian Women's Movement Archives Fonds, X10-1, box 12, folder "Canadian rape crisis centres [Ottawa, ON]: A funding manual for rape crisis centres, 1979." University of Ottawa Archives and Special Collections, Ottawa.

Wigmore, J. (1940). *Evidence in trials at common law* (3rd edition). Boston: Little, Brown and Company.

Wilkerson, A. (1998). Her body her own worst enemy: The medicalization of violence against women. In S. French, W. Teays, & L. M. Purdy (Eds.), *Violence against women: Philosophical perspectives* (pp. 123–135). Ithaca: Cornell University Press.

Williams, C. C., & Williams, R. A. (1973). Rape: A plea for help in the hospital emergency room. *Nursing Forum, 12*(4), 388–401. Medline:4493250

Williams, J., & Holmes, K. (1981). *The second assault: Rape and public attitudes.* Westport, CT: Greenwood Press.

Winner, D. [ca. 1977]. *Rape relief centres.* [Article]. Canadian Women's Movement Archives Fonds, X10-1, box 671, folder "Violence: Day of protest against rape, November 1977." University of Ottawa Archives and Special Collections, Ottawa.

Winner, L. (1980). Do artifacts have politics? *Daedalus, 109*(1), 121–136.

Winner, L. (1986). *The whale and the reactor: A search for limits in an age of high technology.* Chicago: University of Chicago Press.

Wolf, N. (1993). *Fire with fire: The new female power and how to use it.* New York: Fawcett Columbine.

Women Against Violence Against Women. (n.d.). [Press release]. Ann Johnston Fonds, F1312, series 733, file 269. City of Toronto Archives, Toronto.

Women's College Hospital. (1982, November 8). Rape Treatment Centre [Internal memo]. John R. Haslehurst Fonds, HAS, file 73, item 25. The Miss Margaret Robins Archives of Women's College Hospital, Toronto.

Women's College Hospital. (1983, April). Regional sexual assault treatment clinic Women's College Hospital. [Proposal]. John R. Haslehurst Fonds, HAS-73, file 7. The Miss Margaret Robins Archives of Women's College Hospital, Toronto.

Women's College Hospital Public Relations Department. (1982, November). Report on the feasibility study re: Establishing a rape treatment centre at WCH. [Report]. John R. Haslehurst Fonds, HAS, file 73, item 19. The Miss Margaret Robins Archives of Women's College Hospital, Toronto.

Women march in rape protest. (1977, November 4). Toronto Star. Ann Johnston Fonds, F1312, series 733, file 269. City of Toronto Archives, Toronto.

Women rally against rape. (1979, March 8). Newspaper unknown. In RG 23, series G-3, "OPP Crime Prevention Program. Box 14, file "Sexual Assault – Victims" Archives of Ontario. (In Archive Notes: SAEK file, p. 2).

Worster, A., Sardo, A., Thrasher, C., Fernandes, C., & Chemeris, E. (2005, Spring). Understanding the role of nurse practitioners in Canada. *Canadian Journal of Rural Medicine, 10*(2), 89–94. Medline:15842791

Zerbisias, A. (2014). How #BeenRapedNeverReported became a movement. Rabble.ca.

Zook, K. (1980, October/November). Institutionalized rape. *Broadside, 2*(1), 27.

Index

Criminal Code (1990s and 2000s):
consent to DNA samples,
130, 183–4, 196n15, 201n4;
corroborative evidence, 157, 163–4;
DNA typing for other criminal
offences, 128, 194–5n1; law reform
consultations, 105; victim's
personal records, 105–6; victim's
sexual history (rape shield law),
105–6, 127–8, 166–7, 182–3
Crow, Barbara, 138

Davis, Angela, 33
Dawson, Myrna, 105
definitions. *See* terminology
design of SAEK: about, 25–31; actors,
29–31, 186; archival records, 22–3,
55, 191n10; as boundary object,
10–11, 19, 56, 188–9n5; CFS role, 6,
55–6, 120–1; forensic script, 74–5,
77, 81, 84–5, 99–100, 124; history
of science, 20–2, 29; ideal user,
82–3, 93, 151, 154, 167; inscription
of histories, 25–6; marital status,
80; stranger rapist, 80, 86, 123;
translation of victim's experience
into evidence, 32, 56–7, 78, 106–8,
191n11, 195n6; visible injuries,
15, 157–61, 159(f), 190n5. *See also*
SAEK, contents and protocols
design of SAEK (1978): about, 26,
52–5, 62–3; as boundary object,
56; consultations, 55–6, 59–62;
controversies, 83–5; credibility
of victims, 57; first kit (1981),
62–3, 66(f), 73–4, 192n13; historical
background, 45–54; ideal user
(female), 80, 82–3; media coverage,
64, 66(f); objectivity as witness,
56–7, 77–8, 81; power relations,

55–6; protocols, 52–4, 56; PSJ
recommendations, 60–1; testing
of, 57–8; translation of victim's
experience into evidence, 32, 56–7,
78, 191n11, 195n6; victims' reports,
57, 191n11; visible vs invisible
evidence, 57, 80
design of SAEK (1983): about, 26;
resistance to revised kit, 87–90;
time to use kit, 87–8; victims'
reports, 191n11
design of SAEK (2001): about, 102–3,
120–5, 122(f); advocates' role in,
139; benefits of new design, 121–3,
122(f); DNA typing, 26, 101–3, 123,
126–7; samples for DNA testing,
111, 196n11; SANEs' role in, 139;
storage of evidence, 121, 123,
197n22; victims' reports, 191n11;
victim's time to decide to report,
123; working group for revisions,
120–1, 196–7n17
diffraction metaphor, 21–2, 24
Dionne, Martin, 115
disabilities, victims with: access to
SAEK, 149, 194n20; pressure/lack
of pressure to use SAEK, 154–5
dismissal of reports. *See* police,
dismissal of rape reports
DNA Identification Act, 118, 130
DNA typing: about, 26, 101–3,
117–20, 130–1; as actors, 108–11;
automation, 129–30; Bernardo
case, 113–16; compared with
fingerprints, 14, 195–6n9; consent
to DNA samples, 130, 183–4,
196n15, 201n4; controversies, 10,
14, 101–3, 110–14, 125–7, 196n13;
as corroborative evidence, 116–17;
credibility of, 14, 117, 120, 125–6,